INTERVENTION – The Pandora Virus
Copyright 2012 by WRR Munro
Edition October 2013

ISBN: 978-0-9923774-0-3

Cover art
Copyright 2012 by Jeroen Ten Berge

To Emma Munro, without whose advice and encouragement this book would not exist.

And always, to Maya and Chloe.

CONTENTS

INTERVENTION

THE PANDORA VIRUS

A Science Thriller

by

W.R.R. MUNRO

1 :
JUNE 2011

The truth is this: What we are doing is functionally insane. If we do not change this pattern, we will condemn our children and all future generations to struggle with ecological curses for several millennia to come.

...Continuing on our current course would be suicidal for global civilization.

AL GORE, "CLIMATE OF DENIAL," ROLLINGSTONE.COM

Wednesday, June 22nd

He'd nailed it. Had to be right. Maybe he was outgunned, but he had a few tricks up his sleeve.

Cate was on the porch, stretched out on the daybed with her wireless tablet. Marc paused in the doorway, his pulse quickening as he admired her slender, almost feline form. Even after four happy, busy years together, he couldn't quite believe his luck.

"Sorry, Cate. I think I've beaten you to it this time," he said.

She looked up at him and smiled a challenge. "Have you now, my darling?"

"Took a punt on the great apes. Ran my new algorithm against GenBank. Got a 90 percent chance of it being chimp."

A slight frown creased her forehead. "Did you look at the discrepancies? The actual DNA sequences?"

"Hey, I just crunch numbers, remember?"

"Because I initially thought chimp, so I had a look at the FOXP2 transcription factor. I don't think it's chimp," she said.

"The *what* factor? No...don't tell me. Take a look for yourself. It's on-screen now. The desktop." Marc gestured vaguely toward the little study inside. "Nothing else comes close. Coffee?"

"Please."

He busied himself in the kitchen, grinding beans and heating cups, smiling to himself as he created little masterpieces of dense, golden crema and rich aroma. Eventually, he wandered into the study.

"Oh, please no," she said under her breath.

Marc froze.

She sat bolt upright, staring at the screen, her hands held away from the keyboard as if it threatened to burn her. "It can't be."

She must have heard him, smelled the coffee perhaps. She turned around.

"Marc, we're in really big trouble," she said.

"Hey, it can't be that bad. What's going on?"

"It is that bad. It's worse. It's not chimp, Marc. I can tell you exactly what species we have been experimenting with, and believe me, we're in the deepest trouble imaginable."

2 :
APRIL 2033

The wind picked up and the dunes sang. The knife-edge ridge curving along the top of the crescent dune blurred and sand blew into the retreating edge of the rainforest. Small children with dark skin and wide eyes hid amongst the trees and listened to the deep rumbling song that heralded the slow but implacable destruction of everything they knew.

SCIENTIST TV, "AMAZON DESERTIFICATION ACCELERATES," WORLDSCIENTIST.SCI

Tuesday, April 5th

Ayden Walker blinked sweat from his eyes. It was a relief to be working in the relatively dense, scrubby foliage by the riverbed. The cherry trees were pruned hard to reduce irrigation costs through the drought so the orchard offered scant protection from the harsh glare of the sun. Even here though, the air was hot and smelled mainly of dust.

He pushed a thin branch aside and peered through the gloom, just making out the small forms of several bees hovering over wildflowers. He smiled and reached into his pocket for the spray.

[Mark location, image capture,] he commanded silently. As usual, the bees appeared completely normal. There was nothing to indicate that they were failing to pollinate the flowers they visited. Ayden sprayed a fine mist of passive micro-sensors at the bees and slowly worked his sweep net past the branch.

[Marked. Sample 8-323. Still image captured,] his earpiece confirmed.

He swung the net. As its magnetic rim swept past a bee, it momentarily charged the sensors that had come to rest on the insect's surface. Ayden's earpiece chirped, confirming receipt of temperature and humidity data. As delicately as possible, he emptied the net's angry, buzzing contents into a sample jar then crouched and clipped one of the flowers into a sample bag. He rose, moved twenty paces upstream, and was still.

Ayden's earpiece chimed softly.

[Video call from Linsey Carr,] he heard, and a small window appeared in the top right of his specs where it wouldn't block his view. Linsey was sipping from a steaming mug.

[Accept,] he subvocalized, then spoke aloud. "Hey Lin. How's sunny Seattle?"

His audible voice boomed in his ears and shattered the quiet around him. Circumstances permitting, he preferred to speak audibly on person-to-person calls. His phone was quite capable of synthesizing his voice for the listener, and his neckband rarely misinterpreted the minute electrical signals within his larynx when he subvocalized, but somehow the result was thin, lacking timbre. Made him sound like an artificial.

"Oh, the comedian are you? I'll be getting sick of this rain, I'll tell ye that for naught. No video Ayden? Caught you at an inconvenient time, have I?" She peered directly into her cam with a warm smile. Her lilting Scottish accent made her gorgeous West Indian looks all the more surprising, exotic. Married.

He touched thumb and third finger together, reactivating his gesture cuffs, which combined wrist nerve and inertial data to calculate exactly what each finger was doing and integrated seamlessly with the display his specs superimposed over his view. He smiled and flicked Linsey's window larger.

"Nothing so interesting. Just not too many cams around here." He gestured, giving her access to the view from his specs, the line of scraggly trees and bushes, the trickle of water meandering along one edge of the mostly dry riverbed. "Problem?"

"Aye, Ayden, I'm sorry. The replacement RFID sensor spray you need urgently in Montana will be another three, maybe four, days."

"Ahh, okay," Ayden said. He couldn't see any bees in his immediate vicinity. [Mark location, none visible,] he subvocalized.

"George says if you spent more time in the office organizing stuff, we wouldn't have these problems. Says you're not working solo anymore. You're supposed to be managing forty-six field researchers across three states." She rolled her eyes. "He ranted at me for about five minutes—I think you owe me chocolate."

Ayden grinned. As he'd tried to explain to George Reyes more than once, he could access the Bee Anomaly project's virtuality, "BEAN," from anywhere. His boss struggled to grasp that *where* you were just wasn't important anymore.

"One extra creamy with almonds next time I'm in Seattle. It's a promise," he said.

"He says we can't wait, says I should be finding another supplier, but it takes admin a week to approve suppliers in any event," she said.

Ayden's hands danced in the air. Translucent icons floated around him. He shuffled scheduling items. "Don't worry about the spray, Lin," he said. "I can move things around a bit so we don't lose too much time. Just let them know they're pushing it, will you?"

"Okay, but there's one more thing." She hesitated. "George says he thinks MataPharms is already nosing around the EPA."

Ayden's smile vanished. "What are those cowboys up to now? No, let me guess." His voice took on a sardonic edge. "They've engineered a super-bee, resistant to all pathogens, of course… and the honey cures bowel cancer, arthritis, and bad breath, right? And all they want to do is loose their little experiment on the world without all that tedious testing."

She offered him a wry smile. "Something like that, I'm sure."

As she signed off, Ayden shook his head. Then he cleared his specs and filled his lungs, tasting the complexity of smells in the living air. He stood still, allowing the soft sounds of small creatures moving around in the undergrowth to reassert themselves. His smile returned.

• • •

Greg Fanshaw sat down heavily at the big, wooden table that dominated the farmhouse kitchen, holding a couple of fingers to his earpiece as if he could project a tender touch across the airwaves. "Where are you now, hon?"

Her sobbing paused. "I'm in the car. Still in the car park. Oh, Greg, it was just awful. That Jane Hawkins was right behind me. Didn't say a word but she was watching everything."

"Just come home, hon. Are you okay to drive?"

"She'll be online right now, I just know it...telling everyone how I couldn't even pay for the groceries."

"So...what did you do?"

"What could I do? I started taking things out 'til I could cover the bill, that's what I did, Greg. So it's no steak and no beer. We're eating spaghetti and drinking tea tonight."

"It's okay, hon. Just come home now. We'll sort it out."

"It's most definitely *not* okay. I've never been so humiliated in my life."

Greg listened to his wife sob.

"Sorry," she said. "Sorry...I'll be home shortly. Just need to calm down a bit so I don't have an accident."

Greg sat in the kitchen, unable to summon the energy to move. The local hives all had the same problem, and the interstate beekeepers were keeping well away from Oregon. The trees would only be in flower for another week. After that, without pollination, his cherry yields would be too low to cover even the harvesting costs. The bank had already refused to increase their overdraft. He'd stretched it to the limit buying the additional water rights he needed to keep the trees alive through the drought. They'd already sold off Raycliff, nearly a third of the property his father had spent a lifetime building up.

Greg sighed then abruptly scrubbed at his scalp through his grey, closely cropped hair. He reached for the screen sitting on the table in front of him and transferred another thousand dollars from Greg Junior's college fund into the household account. Helen must never find out.

He glanced up at the old side-by-side shotgun hanging on its rack. He had lots of life insurance. It'd only take a careless cleaning accident. "Don't tempt me, old girl. Don't tempt me."

Through sheer, stubborn force of will, he pushed himself to his feet, jammed on his hat and headed out the kitchen door. Where were those useless, damned scientists anyway? He searched through the heat and dust.

Finally, he spotted a lone figure in the shade by the river, standing totally still like one of those street artists. He wore some sort of utility vest draped with little bags and bottles and an insect net, over a crumpled white shirt, sleeves rolled to his elbows. About average height and slim with longish, sandy-colored hair, there was something graceful about him that made Greg feel lumpy. He looked like a long distance runner, a dancer perhaps.

Doing nothing. Greg clenched his teeth as he headed over.

"Mind if I ask what the heck you're doing?"

The kid turned, making a tiny gesture with his right hand. He had all the latest graphene phone gear which, Greg couldn't help thinking, looked like girl's jewelry. It also looked expensive. No doubt Greg Junior would be wanting the same junk in a few years' time, not that they'd be able to afford it. Greg could vaguely see shapes frozen onto the surfaces of the kid's specs. He was probably submerged in 3D porn while he was supposed to be working.

"You must be Greg Fanshaw. G'day, it's nice to—"

"Do ya realize my family's livelihood's at stake here?"

"Mr. Fanshaw, I—"

"The university said they'd send scientists down here to work out what was wrong with the bees. Instead, they send a bunch of kids…and as far as I see you're doing nothing."

"We're doing everything we can, Mr. Fanshaw. We're done collecting specimens in your fields. I'm just finishing up along the river. If you'd like, I can send you our data."

"You're Walker, aren't you?" Greg didn't wait for confirmation. "Well, Ayden, I don't want to know about specimens and such. I want to know when you folks are going to work out what the heck's going on around here."

"Look, I'm sorry, but science doesn't work that way. First we collect the evidence, then we—"

"Don't patronize me, boy. I went to school. Your scientific method's all very well, but my family is sinking without a trace and I need to know what to do about it."

Greg caught the tiny hand gesture, could vaguely see shapes shifting across the kid's lenses. The kid didn't give a damn. He felt himself lean forward, felt his fists bunch.

"I'm sorry but we really are going as fast as we—"

But Greg was already walking away. Had to. To stop himself from hitting the kid.

• • •

Ayden watched the man's rigid, retreating back.

[Found. Proposed emergency financial assistance for primary producers affected by the developing bee crisis,] his earpiece said.

[Memo to project admin. If the financial assistance package gets approved, I suggest we offer it to the Fanshaws. In the meantime, can we look into local community groups? Mr. Fanshaw is showing signs of severe stress.]

Ayden moved another twenty paces along the riverbank.

[Memorandum confirmed.]

He scanned the area for bees. He couldn't think of anything more he could do for the man.

[Mark location, none visible.]

• • •

Back at his campsite, Ayden stacked the last of the day's sample bags in the field fridge, glancing at the data windowed to his specs. The day's strong sun—and his careful alignment of the plastic photovoltaic cells printed onto its lid—had recharged the batteries to 41 percent of capacity, plenty to last overnight and for the trip into town tomorrow. His tent, just big enough to hold his unrolled sleeping bag, was pitched in a small, grassy clearing a few paces from the river bank. He sat next to it, facing the little cams built into the frame of the backpack that held most of his field equipment.

Real video was much better than avatars or just voice when you were trying to be persuasive.

[Video call, Professor Sherman.]

[Confirmed. Awaiting response. Accepted.]

The office appeared to be empty. Much of the flat surface of the professor's desk was occupied by a touch-pad displaying icons and links, and even a virtual keyboard. Professor Hamish Sherman was the head of the Entomology Lab at University of Washington. He was old school, refusing to have anything to do with gesture cuffs, sub-vocalization neckbands, or display specs. He had a binocular cam built into his display so Ayden's specs could display his office in 3D, but he would only see Ayden on a 2D flat-screen.

"Hello, Ayden, my lad. Just one moment please." The voice sounded a little distant, then finally the professor walked into view and sat down with a flourish. As usual, his thin, white hair was immaculately combed, his rich blue shirt perfectly pressed. "Good timing. I was about to call you. George tells me some fool has leaked the vector map to the press and they're all over it. The greenies are having a field day, and California's almond growers are already up in arms. They'll be hit by next summer at the latest. If it spreads nationwide, the financial estimates are in excess of thirty billion per annum, and God only knows what it'll do to food prices. No one has tried to work out the long-term biodiversity impact yet."

"We just can't let it happen, Professor."

"Indubitably. Invectives are being liberally applied in all directions, and our esteemed leader is in damage control mode, but it's not clear how we are going to break the deadlock. I have all but crawled up the backside of every specimen sent to me for analysis, but I can't find anything stopping those bees from doing their jobs. They appear to be perfectly happy and healthy."

The professor's encyclopedic knowledge of all things insect was legendary. If he couldn't find anything wrong with the bees then the project really was in trouble.

"It must be the bees. Too many different plant species are affected. They're all producing normal levels of healthy pollen, and when we pollinate manually, we get normal outcomes."

"You know, Ayden, I was delighted when George approached the university to partner on his pitch to the EPA for this investigation. It's good for the department's budget, and it's good for the students to be involved in something real, but I have to say, I find myself more suited to the pace of theoretical academia. But you didn't call me so we could share our woes, did you? You want something specific."

"Yes, Professor, I do. I want you to come out here, look at the whole picture."

"Ahh, view the whole complex interaction, that kind of thing, is it?"

"Yes."

The professor paused for a few moments. "Well, you know I'm an old-fashioned reductionist myself, but I must admit I can't think of any other approach to take. And, to be frank, the idea of putting a little distance between myself and George does seem attractive at the moment. I'm all yours."

3 :
JUNE 2011

Today humanity uses the equivalent of 1.5 planets to provide the resources we use and absorb our waste. This means it now takes the Earth one year and six months to regenerate what we use in a year.

Moderate U.N. scenarios suggest that if current population and consumption trends continue, by the 2030s, we will need the equivalent of two Earths to support us.

And of course, we only have one.

GLOBAL FOOTPRINT NETWORK, FOOTPRINTNETWORK.ORG

Monday, June 27th

"Marc."

"Hmmm?"

"What are you doing?"

"Just trying to hide your data. You've heard of those algorithms they use to hide data inside photo JPEGs? By encrypting it into subtle color distortions?"

He looked up from the desktop screen. Cate was standing in the study doorway, clutching an open box to the old. paint-stained t-shirt she was wearing. She looked at him blankly.

"Anyway, I think I've found a way to do the same for digital movie files."

"Marc! Please don't bury yourself in some fascinating coding project. We have to get out of here. As far away as we can, as fast as we can, then—"

"Then what? Let him get away with it?"

"No. Maybe. I don't know, but we have a fourteen-month-old baby. Can we just get him out of here first?"

"C'mon, Cate. We're not in physical danger. Gardner may be a bit of a megalomaniac but he's also a scientist. He's dedicated himself to medical advancement, to helping people. He's not a thug. He's not going to—"

"You've no idea what he will or won't do, nor do I. We had no idea how far he'd take our work."

"Look, we'll get out of here today...this afternoon. I'll have this thing running in a few minutes, then I'll help pack. But we need to take the evidence with us. Otherwise, he just has to run a shredder and fire up the incinerator for a few hours and it's our word against his. Okay?"

At first she didn't respond. Marc was about to turn back to his computer, but Cate's expression stopped him. She'd never looked at him like that before. He hoped she wouldn't again.

"What?"

"Evidence, Marc. What would Gardner have to do if he wanted to destroy *all* the evidence?"

4:
APRIL 2033

The overly-handsome young presenter wore a wool suit and scarf but no hat or gloves. Over his shoulder the cloudless sky was an intense purple-blue.

"We're here," he said. "I'm standing at the North Pole. The camera pulled back from the reporter, revealing the white painted metal deck of a ship.

"Of course, I'm also floating at the North Pole because for the first time in human history the Pole is ice free. And that's not all," he said slipping a little thermometer from his jacket pocket.

"It's two degrees above freezing! Now, the whole Arctic isn't ice-free yet. Scientists say that's still a decade away. In fact tomorrow, we're hoping to take you to the edge of the ice sheet and we'll go looking for polar bears! None have been spotted so far this year, but maybe we'll get lucky!

His face became serious. "We have to wait for the all-clear though. There's been another methane event, and the levels are still too high for our safety."

An inset window showed a section of ocean alongside an ice sheet. The ocean was foaming with bubbles.

A SUMMER TO REMEMBER: EPISODE THREE, "CLIMATE CHANGE DELIVERS ANOTHER UNWELCOME MILESTONE AND SCIENTISTS CONFIRM FEARS ABOUT MELTING METHANE CLATHRATES," ASUMMERTOREMEMBER.INDIE.ENT

Friday, April 8th

The orderly rows of white cherry blossoms were starkly beautiful against the landscape of bare dirt and browned grasses. They walked along a row of trees, Ayden carrying Professor Sherman's folding chair. He noted with a smile that the professor wore elastic bands over his shirt sleeves and trousers at wrist and ankle. He wore sunblock under a beekeeper's hat with a built-in veil and was pulling on gloves as they walked across the field. Most of the professor's career had been spent inside labs and libraries.

For no obvious reason, the professor stopped and reached for his chair.

"Thank you, my boy." He unfolded the chair and placed it facing a row of trees.

There was a lot of bee activity at that spot, but Ayden couldn't see how it was different from any other in the field. The professor simply sat and watched, his head no more than a few feet from the activity.

Ayden made himself comfortable under a tree at the edge of the field and slipped into BEAN, setting his specs to 100 percent opacity. The grass and trees and farmland disappeared. Only the deep blue sky remained, above him, below him, infinite in every direction. Emptied of clouds, but not empty. He entered BEAN where he had left it, of course. In front of him was the report he'd been working on, a stack of pages of text closely surrounded by icons for graphs, tabular data, and videos. With a finger tap from one of his translucent, disembodied hands—the only visual indication of his existence—the page he had been working on would dominate his view and he would be ready to continue dictating. It would have to wait, though. He had some housekeeping to do.

He drew back. He wasn't conscious of how he did so, probably leaned back a little, enough for his phone gear to interpret his intention. As he receded from the report, the little group of icons lost opacity and shrank to a single shape, surrounded by many others. Further back, all the report icons appeared as one. Icons for resources which were related to reports were close by. Many more appeared further away and more transparent. Some were small, indicating limited content. Others were huge, representing extensive content. The EPA's faint logo loomed massively in the distance, partially eclipsed by many smaller icons.

His view flashed to the Field Metrics icon cluster, and Ayden quickly drilled down to the latest results from Yakima County in Washington State. The latest results from Yakima had generated a flag. Ayden linked through to the source data.

[Map overlay,] he subvocalized.

The map showed a clear pattern defined by fields and fences, not something bees generally paid much attention to.

[Hold. Overlay researcher allocations.]

It was a clear match. A researcher named Georgina Perez. She reported to Gerald.

[Memo to Gerald Formanu. Hey Gerry, have a look at this (link). Georgina Perez isn't collecting her samples properly. Have another researcher go over her ground, will you? And read her the riot act? She's either got to get serious or—]

Ayden paused the memo. He should have a quick look

[Personal file. Georgina Perez.]

It included a brief note from Gerry. Georgina's aunt had died, but Georgina hadn't wanted any time off. Ayden sighed. He flicked away the last two sentences.

[Show her the data and suggest some time off again. Let me know how it goes.]

Ayden brushed aside the open documents and got back to his interim report. Every so often, he would stand up to stretch and check on the professor. He had to admit that, for a man who didn't like working in the field, Professor Sherman was persistent.

Suddenly, Ayden's specs cleared. Its cams had detected something approaching, and the safety feature had kicked in. It was Professor Sherman.

"Collection kits, insect and plant. Grab them quickly, please. You're right. There is something odd going on here."

Ayden hopped up, headed for their vehicle. "What is it, Professor? What have you seen?"

"No, my lad. I won't give you half-baked guesses based on an old man's failing eyesight. You'll have to wait for the data."

So Ayden and the professor collected bees and flowers from the trees where Sherman had been sitting.

Then Professor Sherman surprised Ayden. He was peering at their samples. "This isn't going to work. Ayden, we're going to need to set up a field lab. Here. I've got to be able to analyze these specimens in situ. Ah, we'll also need a high-speed, high-resolution cam. Can you arrange all that, or do I need to clear it with George?"

"I'll get it sorted, Professor." Ayden flicked open a scheduling window and U Dub's field research asset register. [Video call. Linsey Carr.]

Thursday, April 14th

It took a massive effort. Ayden and Linsey begged, borrowed, and bribed. The following week they had two tents serving as field labs, one as close as possible to clean-room conditions. Professor Sherman, Ayden and three lab assistants slept in spare rooms in Greg Fanshaw's farmhouse. Ayden had managed to convince George Reyes to pay Mr. Fanshaw for food and board. It wasn't a lot of money, but Mr. Fanshaw seemed to appreciate it. There was certainly enough room. Though careworn, the old house had the size and grace of a historic manor.

Ayden sat comfortably on the porch, bathed in the golden glow of twilight, sipping some of Mrs. Fanshaw's endless supply of tea. They'd finally captured some good footage, both at the farm and from an unaffected site in California. Ayden was editing together a three-minute video blog. Next, he'd do a thirty-second version for the media services.

George had insisted they be of the highest quality. "They'll define the public's perception of this entire project. I absolutely must see and approve them before they're posted."

Video window maxed, Ayden watched a split-screen view showing two almost identical scenes. On each, a single bee landed on a blossom, seeking nectar. The views zoomed closer, and the bees bumped against the flower's stamen and pollen fell to the petals. As the bees took off, the images froze and tightened to show the bees' legs. On the left, Ayden could see blobs of pollen. The bee on the right appeared free of any pollen. The scenes morphed—this had taken some dogged editing—so that each showed a single bee leg against the indistinct white background of a microscope slide. Immediately, the views zoomed incredibly close so that individual leg hairs looked like tree trunks. They were virtually identical. The hair on the right, however, had a clearly detectable sheen.

[Send clip to George Reyes. Meta message: if you're happy with this I'll draft a narration and then do the thirty-second edit.]

"Still at it, Ayden?"

Ayden looked up as Professor Sherman bustled onto the veranda, carrying a glass of wine, and settled into the cane lounge opposite his, with a dramatic sigh of contentment. He wore a navy blue blazer sporting a crisp, white pocket handkerchief. All he needed to complete the outfit was a straw hat. Ayden flicked his editing and memo windows aside.

"So, the wax…what is it? Come on, Professor. Give."

Professor Sherman laughed. "You've been most patient, my boy, though I'm sure you've worked it out yourself anyway. As you say, the bees' leg hairs are coated with some sort of waxy substance. That much is obvious. The rest is uncertain at this point, but the substance seems to be related to a bacterium living on the body of the bees. One we've yet to identify. Certainly nothing I've seen before. I've sent specimens back to Watkins at U Dub and we've got your mother examining the DNA. In any case, pollen doesn't stick to the wax."

The professor slapped the back of his neck. "Damned mosquitoes. Should be in a specimen jar where they belong. A couple of degrees of global-sodding-warming and the twice-damned mosquitoes are set to take the world over from the ants."

"The wax isn't visible to the naked eye. What gave it away?"

"Ahh…that's where you were both right and wrong, dear boy. You were wrong about the complex interaction of lots of different factors. Our pollination problem seems to be the straightforward result of a single factor."

Ayden thought it too early to call on that score, at least until they understood where the bacteria came from.

Professor Sherman leaned back into the lounge, ignoring its creak of protest. He was clearly enjoying himself.

"But you were right about me needing to see the bees and plants in their normal environment," he said. "I could have thrashed about in the lab for months before coming across the substance. All I did was look very closely at some flowers, then sit back and watch as numerous bees visited my flowers. Then, I had another look. Clearly the bees were retrieving nectar, and the flower stamen was disturbed as usual, but the pollen was just lying on the petals below—too much of it. And when I looked closely, I couldn't see pollen on the bees' legs as they flew away. You usually can, you know, if you look closely."

"Well, congratulations, Professor. You may have saved the project."

"Thank you for dragging me out here, dear boy. Now, if we're through patting each other on the back, we have a lot more work to do if we are to find out who our bee hitchhiker is. And I think, with your permission, that I will retreat to the comfort of my labs."

Saturday, April 16th

Leonard McFarlane sat back in his ergonomically customized chair, gazing through the view over San Francisco Bay as though it weren't there. He wasn't obese—regular transplants of tailored gut microbes ensured that—but his bulging stomach betrayed his love of eating.

"Next item," he said.

[Possible watch-list flag. Subject of interest: Ayden Walker,] his earpiece said. [Walker is head of field research for the EPA's Bee Anomaly investigation. The investigation has just announced a preliminary finding, blaming an unidentified bacterium for leaving a waxy deposit on the leg hairs of the bees, which affects the ability of the hairs to collect pollen.]

Leonard reached for a ginger biscuit and recalled what he could of Walker, highly intelligent and capable but lacking ambition. He'd radically fast-tracked school and university, by the age of twenty, holding a Ph.D. in complex systems and master's degrees in biological chemistry, and of all things, quantum geometry. But he refused to specialize, making it virtually impossible for him to make tenure anywhere. He had fallen into working as an itinerant field researcher.

It seemed his mother, a respected molecular biologist on the investigation's steering committee and personally close to George Reyes, had wangled a role for her son. Some things never change.

Leonard could imagine that Mr. Hanford might have, at one time, considered employing Walker, but he couldn't see why he would still be of interest. Mr. Hanford had no time for unmotivated employees. He was one of the new breed. Back in the twentieth century, personal power depended on who you could influence, who you knew. Now the dominant players developed and maintained their power because of *what* they knew.

The first few decades of the information age were all about the amount of data that could be captured and made available. Data that would have previously taken months of research to accumulate could be summoned with a few gestures. The problem then became data overload, it became seeing through the mass of data, discerning and absorbing and using the useful information hiding inside.

This was William Hanford's genius.

He was driven and incredibly focused, the opposite of Walker, really. At twenty-four years of age, he was already on the board of directors of the multinational biotech giant Genenco, the youngest by nearly two decades. He was also involved with a bewildering array of research trusts and scientific foundations.

Leonard was one of fifteen senior analysts working in shifts around the clock, seven days a week, filtering and prioritizing information. They were assisted by thirty-four junior analysts who waded through management reports, industry analyses, scientific papers, political commentaries, EPA bulletins, even long-range weather forecasts.

Somehow, Mr. Hanford absorbed it all.

It was borderline, but it was the first time Walker had done anything that might justify a flag.

"Flag it. Level six."

The lowest level.

5:
JULY 2011

The demands of nearly 7 billion humans are stretching Earth to breaking point. We know about climate change, but what about other threats? To what extent do pollution, acidifying oceans, mass extinctions, dead zones in the sea, and other environmental problems really matter? We can't keep stressing these systems indefinitely, but at what point will they bite back?

... However you cut it, our life-support systems are not in good shape.

FRED PEARCE, "EARTH'S NINE LIVES," NEWSCIENTIST.COM

Friday, July 29th

Marc and Cate sat in the living room of a small cottage on half an acre of land, just northwest of Sydney in Australia. Marc watched as Cate tucked a stray wisp of pale brown hair behind her ear. She gave him a weak smile.

"She wouldn't tell me anything. All she'd say was, she was sorry she'd gotten me into it, but it wasn't safe to talk. She said that people who talked...their lives fell apart."

They were speaking quietly, only a little above a whisper, as if it would make a difference.

"We're living out back of nowhere, Cate, on the other side of the planet, unemployed." He waved a hand, vaguely taking in the room. They sat on a worn, rattan lounge with thin cushions that might once have been white. When they first rented the furnished cottage, they thought it looked charmingly rustic. They soon

21

came to realize that it was quietly decaying. "Seems to me our lives have already fallen apart."

"And if we don't do anything, Gardner will keep his experiments going. I know, I know." She paused. "Okay. But first we need to find ourselves a really good lawyer. I'll speak to—"

The doorbell rang. They both jumped.

Marc went to the front door. He hesitated for a moment, wishing it had a peephole, then took a deep breath and opened it. Marc was quite tall and large-framed, if a little soft around the edges. The policeman at the door, though, was huge and there was nothing soft about him. His chest was an expanse of blue dotted with badges and numbers. Beside him, the policewoman was tiny, almost childlike. Behind them was a woman in plain clothes. Their presence was so unexpected that it took Marc several moments to realize he wasn't listening to the policeman.

"I'm sorry, what did you say?" Marc said.

"We're investigating a credible allegation of child abuse. We have a warrant to search your premises and evaluate your child's health," the policeman said.

"That's impossible."

The policewoman stepped forward, unfolding a piece of paper. "I'm afraid not, sir. Please read this search warrant and allow us immediate access."

"No, no, I mean the child abuse thing's impossible. We'd never do anything to hurt our baby."

Cate reached past Marc. "May I?"

The policewoman handed Cate the warrant.

"Come in. You'll see he's fine. This is some sort of mix-up."

After nearly an hour of intrusion, Marc's astonishment had given way to exasperation and then to anger. "This whole thing's crap. The reason you haven't found anything is that it's complete nonsense. Please leave."

"Nevertheless," the lady in plain clothes, the social worker, explained patiently for the second time, "I'm required by law to take your baby in for medical evaluation. You can, of course, accompany us and, if we get the all-clear, you can all be back home this afternoon."

• • •

After four hours of hell, they were allowed to go home.

"He's asleep." Cate sat down at the kitchen table. Her eyes were bloodshot. She looked as if she'd aged three years in the course of the day.

Marc poured her a glass of wine, opened himself a can of beer, and sat opposite with a tired sigh. He took large swigs, trying to blunt the stress that knifed through him. "Thought he'd never stop crying."

"Probably picking up on our tension," Cate said.

"Not just sick of being poked and prodded by strangers for hours on end, then?" He drained the can.

"Who could possibly have accused us of child abuse? We don't even know anybody here. God, I can't remember being this worn out."

The doorbell rang again. They both jumped. Again. It might have been funny in different circumstances.

It was Gardner's bald-headed lawyer from the States. The one who'd looked after all the contract negotiations. They'd called him the Gnome. His expensive, Italian, double-breasted suit did nothing to hide his short, rounded physique, but it set him in stark contrast with Marc's faded t-shirt and scruffy jeans.

Marc gaped at him, his exhausted brain struggling to comprehend the other man's presence.

"Hello, Marc. Sam Jacobson. May I come in?"

"What the hell? What do you want?"

"Well, there's a bit of paperwork to do. We need to finalize your and Cate's resignations."

"You could have posted it."

"You didn't leave a forwarding address, Marc."

"And yet, here you are." Marc felt Cate's presence behind him.

"Here I am. Hello, Cate." The Gnome wore a smug smile.

"Now's not a good time. Any documentation you need signed, just stick in the post." Marc started to shut the door, hoping the lawyer would go away. "Now you've got our address."

"Mr. Gardner was also keen for me to remind you folks about your obligations under the confidentiality agreements you signed. It might be better to talk inside."

"We're aware of our obligations," Cate said from over Marc's shoulder.

"That's good, Cate. So you know it extends to discussions with other ex-employees?"

"You sonofabitch." Marc felt something rise up inside him. Something dark and ugly. His face flushed.

"I must remind you folks that if you breach confidentiality, even chatting with old colleagues, we will prosecute to the full extent of the law."

"Get away from us."

Marc almost had the door closed when the lawyer said quietly, "You wouldn't want any more trouble with the law, now would you, Marc?"

Marc exploded. The door flew open, and he was in the lawyer's face. "It was you. You organized the child abuse thing. Get away from my family! Leave us alone!"

"Child abuse? Now that's a serious offense, Marc." The lawyer backed out their front gate as Marc advanced on him, but he kept smiling. Smugly. Not even pretending to be surprised. "Do you need legal representation? I could ask around for you. I'm sure I could recommend—"

Suddenly the lawyer was on his back on the roadside and Marc was on top of him, his hands full of the lawyer's lapels, shaking hard. Screaming. "Leave us alone! Just leave us alone!"

Marc flew backward through the air, landing hard between his shoulder blades. He was gasping for breath when the huge police officer—*where the hell had he come from?*—drove his knee into Marc's solar plexus. Marc fought for air.

"Easy there, matey," the policeman said mildly. Then, grabbing Marc's shoulder, flipped his bulk over as easy as Marc would flip an egg. He continued conversationally. "Assaulting adults is a lot better than doing it to kids, mate, but we don't really tolerate either around here."

Marc felt handcuffs clench tight on his wrists.

"Now you lay nice an' quiet so I don't have to bounce you down the road into my paddy wagon. Okay?"

Once Marc had mastered the act of breathing again, he lay quietly, only half listening as the tiny policewoman tried to calm Cate and the huge policeman asked the lawyer if he wanted to press charges.

The police. They'd left over an hour ago.

W.R.R. MUNRO

6 :
MAY 2033

Immediately behind the reporter in her spotless sage green suit, a small grubby crowd held up worn-looking placards demanding jobs or more food stamps and chanted half-heartedly, but most of the inhabitants of the tent city, which stretched away in the distance, ignored the protest.

"The riots may be over but Chicago's sit-in is now in its third month. City officials estimate that 150,000 people are involved, and they worry that Grant Park will become a permanent slum. The official figures have stabilized, but protest leaders claim they're misleading. Officially, national unemployment is holding at 14.9 percent and according to the Department of Labor, U.S. households are spending 21.3 percent of disposable income on food, down nearly half a percent from last year's historical highs.

"The protesters point out though, that with record numbers of us working part-time, the Chicago Center for Labor Market Studies estimates that over 18 percent of the employed workforce are unable to get enough hours per week to support their families. They estimate that nearly half the nation spends 35 percent or more of disposable income on food and a fifth spend over 45 percent."

The reporter stepped sideways and waved a hand toward the mass of people, tents, and rubbish behind her. A group of kids were painting dark lines on their faces with a charred stick left over from a cooled cooking fire.

"No wonder parts of the U.S. are starting to look third world."

WSN Newswire, "Grant Park Tent City Becomes a Focal Point for National Anger about the Economy," wsn.news

Monday, May 2nd

George Reyes had booked out Persephone's on Pier 56 in downtown Seattle, bringing together all the members of the core research team and their partners to celebrate. It was the first time the team, or most of it at least, had gathered in one place. He stood with his back to the big picture windows overlooking Puget Sound. Well into his sixties and shorter than average, George Reyes nevertheless cut a fine figure. He obviously kept himself in shape and his collarless shirt and charcoal business suit were the latest in corporate power-wear. His hair didn't reveal a touch of grey, and his round face was lightly tanned.

"I want to thank you all for making the effort to be here tonight. There's lots of work to do yet but, at the risk of tempting fate, I believe we've made critical progress. You're all aware, of course, of the bacterium and its waxy deposits. Now, thanks to the speedy efforts of the DNA sequencing team, I can give you an update. First, we've confirmed it's new to science. Second, and oddly perhaps, it appears to be closely related to an aquatic bacteria, *Pseudomonas Anostraca*, Ps Ano for short. Our bee bacterium, which we've dubbed Ps Novo, appears to have started out as Ps Ano in the water using fairy shrimp as a host then mutated, adapting to life on land. The wax is essentially just waste produced by the bacteria. The details are available on BEAN."

George Reyes smiled widely. "No, don't go looking it up now. Eat, drink, and get to know one another. Enjoy." He waved a hand magnanimously.

A long, cloth-covered table ran down the back wall of the restaurant, adorned with plates of food. Noodle and rice and tofu dishes dominated, of course, but there were also skewers with small portions of chicken and even beef. Ayden's family was relatively well off. His parents managed to get consistent work and Marc was a clever investor but even for them, beef was an expensive indulgence reserved for special occasions.

Ayden and his father were at the back of the loose crowd that surged toward the food table. "More likely some idiot's been playing around with bacterial genomes and failed spectacularly to maintain simple lab procedures," Ayden said with disgust. "Again."

"Not necessarily. High mutation rates are a key survival strategy for bacteria, aren't they? Maybe it migrated from the ocean hundreds of years ago and only recently transferred to the bees from some exotic African insect or something."

"What George hasn't mentioned is there're relatively few single nucleotide differences between Ano and Novo. Just one full gene missing and two extras. This is recent, very likely engineered."

"Maybe, but you're too quick to point the finger at people."

"C'mon, Dad! The inmates are running the asylum. What do you bet some idiot was trying to 'improve' the bees?"

"Hang on a sec, Ayden," his father said. [Your mother's in danger of terminal boredom and has demanded that I rescue her immediately,] he subvocalized so that no one would overhear.

"Okay. See you," Ayden said as his father turned away. [Ask her. No way is this thing natural,] he subvocalized.

[Okay, okay. I will.]

A "Waiter's Friend" trolley stopped in front of him. It tried to message Ayden, no doubt an invitation to order a drink. He ignored the message and gave the nudge-sensor bar near its base a gentle kick. The trolley's programming was simple and deliberately timid—it was, after all, designed to serve drinks to people in various stages of inebriation. A bump meant it was in the way and should move off in an unblocked direction. It did so, pausing to allow its infra-red sensors to plot dense groupings of people in the room and head toward one of them chosen at random.

He was about to go looking for Professor Sherman when he found himself facing a fresh-faced girl, a little less than medium height with unruly strawberry blonde hair and just a hint of freckles. And strikingly large, round eyes. She was quite beautiful in an understated sort of way, so of course, Ayden wasn't sure what to say.

"Hi, Boss," she said. "Tess James. One of your U Dub under-grad slave laborers."

He couldn't decide if they were blue or green or perhaps even silver. "Hmmm…that's possibly the first time I've been called "boss." I think I find it a little uncomfortable, especially since we're all around the same age. But I'm not really anyone's boss. I'm just the field researcher who got lumped with sorting out all the logistical messes." He glanced at the tag hovering in his specs over her left shoulder. "Ahh, you're with the Montana field team. Nice to meet you."

"Likewise. I have a question. I looked up your CV. Doesn't it worry you, not having a career path?" A playful smile lit her face.

"Ouch. You don't pull your punches, do you?"

She said nothing, her smile broadening.

"I get involved in interesting jobs," Ayden said. "I guess that's all I really want right now."

"Okay, but what about money, security, all that stuff?"

"Well…I figure I'll keep making myself useful and earn enough money to put food on the table, without becoming a slave to it. I am buying a condo in San Francisco, may even be able to afford to live in it one day."

"That's scarily rational. Everyone I know is terrified of joining the unemployment queues. Is that an Aussie thing?"

"I grew up in Australia, but actually I'm a Yank."

"Hah, no you're not. Yankees are from the north, unless you're from there, then they're from the northeast or from Vermont or something. No one in America calls themselves a Yank."

"Just one of my many bad habits, I guess." Ayden smiled. "You know Aussies use it as a term of endearment, right?"

"Really? So what's with BioWatch? They're anti-science, aren't they? Says here you're one of their leaders," she said with a grin. She was definitely teasing him.

Ayden laughed. "That's not on my CV. What are you looking at?"

She returned his laugh and flicked a finger. In milliseconds, her phone combined her cuff data with the visual pickup from her specs to determine whom the intended target was, used her triangulated location to calculate his, performed a public lookup, negotiated a secure communications protocol with Ayden's phone and offered him a link. [No one actually trusts CVs do they?]

[I suppose not.] He flicked open the blog link. "Seems not everyone likes us. [Surprise, surprise.] Really though, I'm not a leader. BioWatch isn't organized enough to need leaders. It's really just a forum for people to compare notes and raise concerns. Mainly I try to increase awareness about specific threats to essential ecosystem services. And most of those come from unsustainable consumption rather than anything science produces."

"Okay, but according to this, you're on record attacking the big transgene companies."

"Well, I'm also against commercializing stuff before it's understood. They release new genetically modified organisms into the environment with barely a thought."

"The EPA approves every GMO before it's licensed, doesn't it?"

"Sure, after a one-dimensional look at the risks, but they don't take complexity into account. It's like their approval of chemicals last century. 'Safe' exposure levels were set, completely ignoring the rest of the cocktail of chemicals in the environment. Remember the Cheng Report?"

"Umm, no."

"Millions of deaths from breast cancer attributed to the additive effects of xenoestrogens found in different commercial products, all at approved safe levels, until they were mixed together, that is."

"Yeah, but now regulatory approval has been tightened up, hasn't it? So much that industry whines that we miss out on all sorts of product improvements because they can't justify funding the approval process?"

Her dismissive expression made it clear she felt little sympathy for them.

"In the chemicals industry sure," Ayden said. "But biotech is an order of magnitude more messy. GMOs are analyzed for their interactions with associated species, largely ignoring how they might mutate and interact with unrelated organisms. That's pretty dumb when microorganisms swap DNA on a regular basis."

"Horizontal transfer, yeah, but factoring in all the possible permutations would be impossible, wouldn't it?"

Tess snagged a couple of canapés from a passing Waiter's Friend and offered Ayden a spring roll.

"Not impossible, but expensive, I'll grant you." He took a bite. "That's really good."

She nodded, her mouth full. The crowd had broken into small conversational groups plus a few lone undergraduates hungrily pursuing Waiter's Friends around the room.

"So we say 'to hell with the risk, let's just do it blind,'" Ayden said. "I mean, curing a cancer justifies some risk, sure, but personally I'd prefer to miss out on the latest improvement in pet shampoo to avoid risking irreversible damage to the food web, especially…"

Ayden's voice trailed off. Tess's large eyes widened even more. She looked like she was about to start bouncing up and down with excitement. At that moment, he couldn't have taken his eyes off her if he'd tried. "What?" he asked.

"Wait, um, okay, back to xenoestrogen cocktails for a minute. You may have hit on the answer to a mystery I've been struggling with."

Ayden raised an eyebrow.

"Well, it's to do with male fertility. Strictly speaking, it's outside my specialty, but I guess you'd approve of that." Her voice sped up. "My cousin, Nikki, married her dream man, her Mr. Right, but it turns out he's got a fertility problem. Anyway, they've ended up going with IVF and they have beautiful baby twin girls so it's all good now, but the thing is, his sperm appeared perfect. His count was high, motility was good, and morphological deformity was low. Once they had fertilized some of her eggs, they implanted first time, which often doesn't happen with IVF. His sperm just couldn't penetrate her oocyte membrane. Apparently, it's a capacitation problem."

This was a lot more detail about her family's private affairs than Ayden was expecting. "I'm not familiar with…"

"It's got to do with the removal of a glycoprotein layer and alteration of the aerosomal cap. Basically, it's the sperm preparing itself for penetration of the egg while traveling through the female genital tract."

"But his doesn't, right?"

"Yeah, and Craig sees it as some sort of personal failure." She smiled. "I mean, he's your typical gung-ho overachiever, a jock with a brain. College football, Harvard law, all that."

She paused, her face becoming totally serious.

"Thing is, these days you can rely on the medical profession to offer endless reassuring statistics. You expect them to say 'don't worry, sir, one in twelve males has this problem' or something like that, but in this case, they don't have the figures. They say capacitation problems are increasingly common, but they don't know how long it's been common for or just how common it is. It looks like the medical community has been caught napping. Part of the problem is, it's not obvious to detect."

"So the data you need hasn't been recorded?"

"I don't know. The IVF companies are being secretive. They refuse to show me their data. But maybe back at the turn of the century, Craig's mother, and a lot of other mothers, were exposed to a combination of chemicals which affected the development of their unborn children. Maybe it's only now that the next generation is trying to breed that the effects are being felt. Maybe this is signaling a new wave of fertility problems."

"That's a lot of maybes. Why don't you flick me what data you do have? I'd be happy to take a look. Might be able to make some suggestions."

"Thanks, Ayden, I'd appreciate it." She beamed at him.

Tuesday, May 3rd

The next morning, Ayden sat soaking up the sun on the postage stamp–sized balcony of the rented condominium he was sharing with his parents for the duration of the project. It was a rare morning in Seattle, not a cloud in the sky, and Ayden was determined to take advantage of it. It had rained overnight, as usual, so the balcony was still damp and the air was steamy. It was hard to believe that much of the Pacific Northwest was officially in its sixth year of drought.

There was, of course, a lot more field work to be done, but the real progress would now be made in the lab. The EPA's terms of reference were actually quite narrow. Find the cause, measure the extent, and project future spread. They weren't tasked with deriving a solution and, strictly speaking, they weren't expected to discover the source of the bacteria, but how could he not try?

Specs set to opaque, Ayden's hands flashed through the air. He searched for *Pseudomonas Anostraca* together with the proteins coded for by Ps Novo's new genes. There were only a dozen or so hits, a *Nature* article and a collection of U.S. Patent Office notices.

The *Nature* article was published on April 2, 2028 and described a paratransgenic *Toxorhynchites Rutilus* mosquito, suggesting its use as a biological control. The mosquitoes themselves were not transgenic; their genomes had been left untouched. They had, however, been infected with a transgenic parasite, a bacteria, hence paratransgenic.

Toxorhynchites larvae preyed on the larvae of other mosquito species, but the adults weren't bloodsuckers so they'd first been proposed as biological controls for bloodsucking mosquitoes in 1911, but this fearsome predator had critical weaknesses. It was a relatively slow breeder that didn't survive dry conditions well.

But Dr. Winston Lowe figured he'd found a solution to those weaknesses and raced to commercialize his success. Ayden shook his head.

It had been known for decades that fairy shrimp eggs could survive out of the water for extended periods of time without drying out. Dr. Lowe had identified that the eggs were crawling with Ps Ano bacteria and that the bacteria assisted the eggs to be desiccation-resistant. So he modified the bacteria to suit its new host, to need it. The waxy substance excreted by the bacteria made the mosquito eggs not only desiccation-resistant, it made them more robust.

Ayden wondered if Lowe had already engaged an advertising agency when he filed his patent application.

"Rhino-MozzieTM eggs will be packaged, stored, transported, and sold as easily as a packet of flour," it read. "Application will be as simple as tossing a pinch of powder into bodies of still water. Large-scale application will be trialed using crop-dusters. The eggs will hatch their carnivorous larvae within a couple of days of application."

Curiously, there were no follow-up articles and no record of approvals being sought or obtained. Dr. Lowe and his patented mosquitoes disappeared from sight. The Patent Office notices revealed only that the patent had been sold to an organization called MTC Trust Incorporated, whose registered office turned out to be that of a legal firm in Delaware. There were no other contact details listed.

[Lookup, MTC Trust Inc., Delaware,] he subvocalized.

[Found,] his earpiece said.

[Voice call.]

[Connecting.] A small window popped into Ayden's specs showing an expensive-looking receptionist fronting an expensive-looking reception.

"Thank you for calling Tollhurst, Thompson, and Rathbone. How may I help you?" she said.

"Could I talk to someone about MTC Trust, please?"

"I'm sorry, sir, no one is available at the moment. Would you care to leave a video message, or would you prefer to correspond by text?"

Ayden flicked the window larger. The receptionist was super-model beautiful in a flawless, anonymous way. Her diction was impeccable. An artificial, for sure. Ayden groaned.

"Can you tell me who, at your firm, is responsible for MTC Trust?"

"The firm of Tollhurst, Thompson, and Rathbone represents MTC Trust Incorporated in all matters, sir. In addition, our managing partner, Mr. Benjamin Tollhurst, is an officer of MTC Trust Incorporated."

"Can I speak to Mr. Tollhurst, please?"

"I'm afraid he is unavailable. Please use this link to leave a message or send a letter. [Link.]"

"Can you tell me when he's likely to become available?"

"No, sir, I cannot. I'm sorry. Mr. Tollhurst is a very busy man."

"Okay. Please ask Mr. Tollhurst to contact me."

"I would be happy to pass on your request, sir. Would you care to tell me the nature of your inquiry?"

There was no point in arguing with an artificial. Infinitely patient, maddeningly polite, the virtual receptionist would run him through the same set of response routines all day. "It's with regard to patent number 12,825,247, purchased from Dr. Winston Lowe in 2029."

Wednesday, May 4th

The following day, with no response, Ayden tried again. The virtual receptionist made it clear that she had discharged her responsibilities in passing the message on, and offered the link again.

Ayden decided to up the ante. He drafted a letter. He described his concerns that an organism, for which MTC Trust Inc. held the patent, may possibly have transferred in an uncontrolled and unapproved manner to wild bees and may be responsible for extensive environmental and commercial damage. Ayden asked for their assistance in investigating these concerns.

He reviewed the letter carefully before sending it. He was, after all, sending it to a law firm.

● ● ●

Three hours later, Ayden's phone chimed.

"What do you think you're doing, Ayden?" George Reyes said. His round face loomed large in Ayden's specs' window, creased with intensity. "I'm looking at an injunction here, requiring that we cease and desist making any allegations relating to MTC Trust Inc., whoever the hell they are, and some patent I don't know anything about. I'm also looking at a court order to hand over copies of any and all documentation relating to those allegations. And a cover letter from some legal firm in Delaware strongly suggests that I give careful consideration to the continued employment of one Ayden Walker. Now, this,"

George waved an arm and two more windows appeared in Ayden's specs.

"is Lachlan Murray, a senior partner at Hollingdale's and an old friend of mine, and his associate Tim Baird. Gentlemen, this is Ayden Walker, who until today has been an exemplary employee."

Murray nodded, his craggy face expressionless. Baird, who looked about the same age as Ayden, smiled a greeting.

"Now, Ayden, I need you to tell us everything about your contact with MTC Trust. Leave nothing out."

Ayden gave them a very quick rundown of his initial suspicions about the mosquitoes, then a detailed description of his conversation with and letter to MTC Trust.

"That's it?" Murray said.

"That's everything."

"We're going to need to see a copy of that letter."

Without a word, Ayden flicked the letter into their shared space.

In all three windows, the men gestured for the document and their eyes dipped as they began reading.

"This is the only thing you sent them?" Murray said.

"Yep."

"And on the phone all you said was you wanted to discuss the patent?"

"That's it. The closest I came to making any allegations is what's in that letter."

"Well, then, I don't think we have a major problem," Murray said to George. "His language is pretty careful."

"Can you make it go away?" George said.

"Probably. My guess is these guys are just sending a message to butt out. If we make it clear we are happy to do that, I suspect they won't pursue it."

"Sir?" Baird said, his eyes still locked on the document. "Can I ask Walker a question?"

"Fire away."

Baird looked up. "Mr. Walker, is this timestamp correct? You only sent this letter at 1:30 this afternoon?"

"That'd be right."

"What?" George said. "Wait a minute…they got a judge to sign a court order less than three hours after receiving the letter? Who the hell are these guys?"

"Good question." Murray said. "We have any answers to that one yet?" he asked Baird.

"One moment, sir." Baird's focus shifted, and his hands began twitching.

Murray looked back to George "This is just a bit scary. These guys must have some clout."

"Certainly looks that way."

"Well if this young man is being totally candid and not forgetting to tell us anything, then I still think we're okay, but clearly they're capable of being a pain in the tail if they want to."

"Ayden," George fixed Ayden with his gaze. "Are you telling us everything we need to know about this?"

"Yes, Dr. Reyes. I've had no other contact with MTC or conversations about them."

"Hmmm. Baird? Anything?"

"Maybe. MTC Trust's ownership is obscure. Looks like a private, offshore trust. A few weeks of drilling may tell us something or it may not. Their lawyers, though, they may give us a clue. According to our database, the only time we've come across them before, they were representing an entity associated with Genenco."

"Genenco?" George shook his head. "Still think we're okay, Lachlan?"

"That does put a different light on the matter. If it is Genenco, they can cause us a world of pain. Their reaction is way over the top, though. It still looks like a message to me. How badly do you want it to go away?" Murray eyed George. "Enough to make a gesture?" Ayden noticed that Murray couldn't help but glance in his direction for the briefest moment.

George just looked at Ayden.

"If by "gesture," Mr. Murray means I've got to go, then…fair enough," Ayden said to George. "I did start this thing."

"Yes, Ayden you did. And if it proves necessary, you should know that I will drop you like a stone. But," George said, turning his gaze back to Murray, "I don't like giving in to corporate thuggery that easily. This is probably just some pissant Delaware lawyer throwing his weight around and thinking he'll get away with it because he's got Genenco behind him. Have a go at smoothing it over for me, will you? Tell them we'll leave them alone but don't offer them anything in particular. I happen to know one or two people with contacts at Genenco. If need be, I'll get in touch with my contacts and see what the story is with MTC, but I'd rather avoid doing that unless necessary."

"No problem, George. I'll draft up a letter. Want to see it before it goes?"

"No, Lachlan, not necessary. Now are you in Seattle? Dinner? The Club?"

"Delighted. I just need an hour to get this happening."

"Perfect, because I need a few minutes with young Mr. Walker here."

Murray smiled as his and Baird's windows disappeared.

Reyes stood and began pacing. At first, his office cam failed to keep him in frame. Then it zoomed out, giving Ayden a wider view of the office, but Reyes was pacing back and forth in front of floor-to-ceiling windows so all Ayden could really see against the glare was Reyes' silhouette.

"Command: dictation. Memo to Ayden Walker. Dated and time stamped. You are hereby informed that you are on leave, starting right now, for an indefinite time period and until explicitly notified otherwise by me, in writing. You are instructed to make no further contact with any external organizations or individuals referencing this project or your role in it. You are notified that if you ignore that instruction and incur any legal action against the project, the project will hold you personally liable for all legal costs and any damages that may result. Command: end dictation. Copy to BEAN notice board. Copy to Walker. Got it, Ayden?"

"Got it."

"Good. Now you can keep in contact internally if you want, but outside this project, I want your lips sealed. I don't think you did anything terribly wrong, just foolish, but you need to lay low for a while 'til this thing blows over. Are we agreed?"

"Yes, Dr. Reyes."

"Good. Now go do whatever it is you do when you're not wandering around the countryside and firing broadsides at multinationals. I'm going out to dinner with an old friend so we can lament the shortcomings of the younger generation."

• • •

Leonard McFarlane leaned forward in his chair, his biscuit forgotten. Ayden Walker had linked the bee bacteria to a patent owned by a Genenco entity. Their lawyers had, of course, fired a warning shot, and George Reyes had backed Walker down.

Leonard flagged it as a level-three priority and immediately started calling up as much detail as possible about Walker's current activities.

• • •

William Hanford contemplated McFarlane's report. Ayden Walker was making broad enquiries and starting to source mosquitoes for analysis. No doubt, he'd soon learn that common blood-sucking mosquitoes infected with the bacteria were widespread.

It was time for William and Ayden to meet.

Time for Ayden to learn the truth about himself.

Friday, May 6th

Ayden was a few minutes early. He gazed around the enormous room, taking in the sea of pale sandstone, the island of deep, warm orange rug, the sumptuous chocolate leather armchairs facing each other across a highly polished, low, wooden table. Soft, golden light bathed the room, courtesy of the beautiful, twilight sky visible through the surrounding windows.

Clearly, William Hanford favored a warm atmosphere over realism for their meeting. It was noon in Seattle, and Ayden understood that William was in the same time zone, down in San Francisco.

Ayden had been intrigued when William messaged him, claiming to have called off MTC's lawyers. Perhaps he would get some straight answers about the mosquitoes after all. At William's insistence, Ayden made his way to the Radisson Hotel, which boasted fully immersive conference rooms. High-resolution cams captured Ayden from every angle and the "wall screens" (really a misnomer since they also covered the floor and ceiling) presented imagery consistent with his specs so that even his peripheral vision, which couldn't see in 3D anyway, was fooled. Prior to the room's virtual reality feed firing up, Ayden had been sitting in a typically small hotel conference room with comfortable but bland décor and no windows.

[Conference call from William Hanford,] his earpiece said.

[Accept.]

Ayden stood as William appeared near the door.

Unquestionably, Ayden had never met William before, but there was something oddly familiar about him. At first, neither man spoke. They simply stood, observing each other. William was tall and angular. He wore a collarless white shirt, buttoned to the neck, under a conservative charcoal suit. His hair was short and very dark, possibly pitch black. Ayden studied his quiet, unlined face. It looked neither young nor old.

The moment stretched.

"You feel it, too," William said.

"Hmm, but it's not as puzzling for you as it is for me," Ayden said.

"I appear to have you at a disadvantage. I thought as much."

Finally, William moved, crossing the room to sit opposite Ayden. Immediately, Ayden understood what it was that generated the feeling of recognition. He had been still. He hadn't shifted weight or fidgeted, nor had he held himself rigid, the way most people do when they are consciously trying not to move. He simply stood comfortably but completely still. Even when he crossed the room and sat down, William had displayed rare economy of movement. Ayden had long since noticed that, while he was comfortable being still, most people seemed incapable of it. It was almost as if homo sapiens weren't designed to stop moving.

"I take it you didn't tell your parents you were meeting me. Pity. If you had, our conversation could have been much more interesting."

"My parents don't usually keep things from me, so they must have had a reason not to mention you. Are we related in some way? What is it you want to tell me?"

"I'm not the subject you need to discuss with them. You are. There's something they should have told you long before now. Crucial information they've been denying you. About who and what you really are."

"*What* I am. What are you talking about?"

"It has to come from them."

Ayden waited, but William obviously didn't need to fill the silence.

"That's it? I came across town for that?"

"The sense of recognition you felt when we met was your subconscious perception of a pattern, a commonality between us. I couldn't be sure that would happen without high-res 3D. Now, you'll force your parents to explain."

"Okay. Why don't I ask them right now?"

"I assure you, it won't be a short conversation or one you'll want to have with an audience. Talk to them in person. When you hear what they have to say, you'll be glad you did."

"Okay, my turn. Did MTC Trust deliberately add a gene to Dr. Lowe's Ps Novo so it could escape Toxorhynchites, and how did it become so widely dispersed?"

"Sorry, Ayden. I really don't want to withhold information from you. In fact, I'm hoping we can become valuable allies, that we can work together toward some critically important goals. But first, you must talk to your parents."

Ayden let some of his frustration enter his voice. "That's not good enough, mate. Someone has released a genetically modified organism into the environment. That organism is causing—has already caused—substantial damage to the ecosystem and many millions of dollars of crop losses, soon to be billions. If your organization is involved, surely you need to think about damage control. Surely, the faster you cooperate, the better it will be for your organization, as well as for agriculture and the environment."

"Talk to them. Then contact me. I'll meet with you again immediately. One final suggestion, if I may? Shut down your phone before you start. Your parents' too."

"Isn't that a bit over the top? All comms are encrypted."

"Complete shutdown, not just privacy mode. It's not the actual comms data that's vulnerable. Don't you ever wonder why exploits are discovered for so many apps? Often after they've been popular for years? Almost makes you wonder if someone has known about the holes all along, doesn't it? Deliberately built them in from day one, even?"

They looked at each other in silence.

There were, of course, endless conspiracy theories about government agencies tapping into people's phones, but Ayden's life had always been an open book. He didn't have anything to hide.

• • •

Marc Walker paced back and forth across the end of the little condominium's living room, brushing past some of the ferns and cycads that Cate cleverly used to counter the beige blandness of its "designer décor."

"You've gotta understand, Ayden, it all seemed so innocent at the time. We wanted kids and weren't having any luck. Then Gardner offered us free IVF and suddenly we had twenty-two viable embryos. Why throw a dice, we thought, when we were world experts?" Marc said.

"Embryo selection?" Ayden said.

"Yes."

"Well, for a start, it was both unethical and illegal."

"Technically, yes," Cate said. "But it was standard practice to screen for all sorts of single-gene defects like cystic fibrosis and muscular dystrophy, at least for high-risk parents. And more and more people were quietly screening for cancer and Alzheimer markers and all sorts of things. We just took it one step further. At least, that's all we thought we were doing."

"We trusted Gardner," Marc said. "I mean, we thought we were in heaven. We were doing seriously fascinating science, we were in love and we were gonna have a baby. And we were in this amazing place! We lived in these little bungalows surrounded by grass and trees, and had a community center with a café and daycare and a preschool.

"We called it Shangri La. Seriously. The original Sanderson Hospital was in the hills outside Fairmont on a huge piece of land. Gardner built the institute within the grounds, and because of commercial secrecy, and the right-to-lifers and the animal-lib crowd, there were these big fences and armed guards around the outside. So inside, kids could run around all over the place completely safe. Inside the fence, there was no unemployment, no crime, no violence, and there was a state-of-the-art hospital on campus. When Gardner suggested screening, it seemed like a no-brainer."

"Okay, you were naïve and optimistic. I get that. But this was decades ago. Most of the research was focused on single nucleotide traits, which account for, what, a few percent of intelligence?"

"Around 4 percent, out of roughly 50 percent regarded as genetic," Cate said.

"So the other 45 percent emerges from the complexity of lots of different genes switching on and off as neurons are created, migrate all over the brain, establish pathways and are pruned, and all this not only within the womb but through the organism's…" Ayden's voice trailed off for a moment.

"Through *my* lifetime. Which means the protein coding sequences wouldn't have been enough. You would've have had to include regulatory code as well. Genes for RNA interference."

"Yes," Cate said. "In fact, most of our work focused on regulatory processes. Smart people and dumb people are made up of the same protein soup. Actually, it was even worse than that. In the end, we had to look beyond DNA."

"Epigenetics?"

"Yes, methyl groups proved to be important. Both trans-generational and environmental."

"But…well, sorry, but how could you have been so arrogant?"

Cate winced and closed her eyes. "Good question."

Marc stopped pacing and locked eyes with Ayden.

"You're used to small science, Ayden. Subtle science. Sanderson Health was more along the lines of Los Alamos. A private sector, biotech version. Cate and I were a tiny part of a huge machine. Gardner had turned a decent inheritance into serious money, hundreds of millions at least, and it was all geared toward his research. He bought the existing hospital and set up the institute, but that wasn't it by a long shot. He had a worldwide network of clinics and some very high-profile successes—everyone wanted to work with him."

He fell silent, staring into the past.

Cate looked up. "So, as well as everything going on within the center, Sanderson's was receiving information from projects all around the world. We were cross-referencing DNA sequences from all sorts of organisms and correlating these with measures of neural complexity and problem-solving ability. The data was inconsistent and contradictory, and there was a lot of it. Marc's nonlinear dynamical systems background was just what we needed."

"We were a sort of 'clearing house' for all the information," Marc said, starting to pace again. "And I've gotta say it was a blast. This was pre-grid. You couldn't just rent whatever processing power you needed in those days. We set up our own mini-grid. We had a bank of a hundred and fifty dedicated PCs, plus the idle capacity of every workstation in the research institute and hospital, and from associated research facilities and fertility clinics all over the world. If I had to get over some impasse, I'd just devise huge, ugly simulations. I could bash my way through problems where finesse and clever math failed."

"We worked as a team," Cate said. "It was how we met. I was a bridge between Marc's mathematical analysis and the genotype and phenotype data that poured in." She stood and smoothed her shirt nervously. "I'm going to make tea," she said.

"Yeah, well actually, I was just a tool Cate used to try to detect patterns," Marc said.

"Okay, but I can access that sort of processing power right now, through my phone, if I want to spend the money, but even given that, and the last twenty years of research, what you were trying to do would be a real challenge today. How could you have hoped to do it then?"

"We couldn't have done it using traditional genetic analysis. It should have taken decades, would have, but for your father's stroke of genius," Cate said from the other side of the kitchen counter that ran along one end of the living room.

"Actually, I didn't tell anyone at the time, but I got the idea from communications engineering. They were using something they called belief propagation. It caught my attention because it posed a fascinating puzzle. The technique worked, worked really well, but no one had been able to provide a mathematical proof for why. I used to play with ideas for a proof but never got anywhere. Then one morning, I woke up knowing I had to try it on our DNA data. In the 1960s, some guy named Gallager proposed what he called a 'low-density parity check code.' They didn't have the computer grunt to try it back then, but by the turn of the century, they had plenty and were using it to massively improve satellite signal-to-noise ratios. I had processing power to burn, so I thought I'd give it a go. I had no idea how well it would work."

"So how does it work?" Ayden said.

"In English, please, Marc," Cate said.

"Umm, basically it involves establishing a whole lot of nodes, each of which analyzes an information string, and applies rules relating to the type of information expected, and assigns probabilities. In the case of satellite interference, probabilities were assigned as to whether each bit of information was corrupted or not. Other nodes would analyze strings partly overlapping the string of the first node, adding their 'vote' to the likelihood of characters being uncorrupted. I applied the approach to DNA. Each of my nodes estimated the likelihood of a sequence being associated with a particular trait. I didn't worry about whether that sequence was solely responsible or simply one of many interacting to produce the trait. I certainly didn't give a thought to the mechanism. Each experiment would yield another set of data to plug in, adding another vote to the probability. The real power of the approach was that it was flexible enough to evaluate inconsistent data from different sources."

"Marc's analysis could even cross species. Data from mice, pigs, monkeys, bats, almost any organism, could be incorporated," Cate said.

"In other words, the reason we were so arrogant is we were succeeding. We were Gardner's star team. He was actively experimenting, applying our analyses, more often than not, with astounding results. Gardner may not have felt bound by ethics or rigorous scientific method, but he sure believed in measuring results. We were responsible for the most intelligent mice, spider monkeys and chimps that ever existed."

"And we looked at human data all the time, from the IVF clinics. We had access to genotype data from the parents and used demographic information as a rough proxy for intelligence where there were no IQ scores. The human data provided more information for Marc's nodes to assess."

"So, you figured you had the knowledge to select the embryo that was likely to produce the most intelligent human child."

His parents had become more animated as they'd lost themselves in the science, but with Ayden's question they seemed to shrink into themselves.

"Yes," Cate said in a small voice. "At least we figured our odds were better than random chance."

After a very long silence, Ayden said, "That's not all though, is it? There's more."

Neither of his parents would meet his eyes.

Ayden waited.

"Yeah," Marc said, eventually. "There's more."

"We think there's more," Cate said. "We can't really prove it." She put the tea cups on the low table and sat beside Ayden on the couch, clutching one of his hands in hers.

As the story unfolded, a small part of Ayden had known how it would end, a small voice he had been intent on ignoring. "Just bloody-well tell me!"

"Okay! Okay." Cate took a deep breath. "The first IVF round didn't work. Months went by. We're pretty sure Gardner didn't limit himself to screening your embryo. We think he, oh God, he interfered with you. He manipulated your DNA to enhance your intelligence. He'd started experimenting with people. You were one of our test organisms."

Ayden stared at her.

"I'm a GMO." His voice sounded strange in his ears, distant. He could feel the blood draining from his face.

Tears streamed down Cate's face. "At first, we weren't sure. We were afraid to find out. The embryo screening that we were knowingly involved in was illegal, and it'd be hard for us to prove we didn't know anything about the rest. We thought we'd lose you, Ayden."

She reached for his other hand.

"We thought they'd take you away from us. We were so afraid, sweetheart. Afraid for you. You don't know what it's like, being a parent." Cate broke off.

Arms around his wife, Marc continued for her. "It's not like we didn't think of exposing him. We planned to. We changed our plans. We went round and round in goddamn circles. But Gardner was watching us, following us. No matter where we went, his people would be there. They made it perfectly clear, if we talked, they'd destroy us and they wouldn't hesitate to ruin your life as well."

"What about after Gardner died? Couldn't you have done something then?"

"We didn't know who else was involved! And we were afraid you'd never be allowed to lead a normal life, that you'd be treated like a laboratory animal."

"A lab rat." He jerked his hands away from hers. "You cut and ran. You didn't tell anyone."

"Our silence bought you a normal life."

"And you figured you could just get on with your lives as if it never happened?" His voice was hard and cold.

"Get on with our lives? Ayden, we've spent the last twenty years afraid of our own shadows. Your mother could have been a world leader in her field. Instead, she's spent two decades applying a tenth of her expertise to crappy little freelance environmental cleanups."

Cate flinched, and Marc turned to her. "Sorry. I didn't mean it like that. I know your work's been important, but where your career could have gone…"

Marc turned back to Ayden. "Mine, too," he said. "My work was pretty ground-breaking. It could have established my career, but I've been afraid to have anything to do with it, so, instead I'm just a statistician for hire."

"To hell with your career! You didn't tell *me*, Dad! You could have told *me*!"

"We planned to, once you grew up, but the time was never right, and then you became so anti-GM."

Ayden didn't say anything for a long time.

"Hey, talk to us, Ayden," Marc said quietly.

"I'm a GMO. I'm a genetically modified bloody organism. What am I supposed to say? What am I supposed to think? Hey, maybe I should try to find a self-help group for people who find out they're GMOs… oh… there are others. There are lots of others. There must be GM humans wandering around all over the place. William. He's like me, isn't he? He's GM, too."

"Yeah, we think so. Maybe." Marc said. "One of the other kids at Sanderson's was called Will Hanford. A year or two older than you. Truth is, we don't know for sure. We couldn't risk trying to find out. If we're right, then…yes, there are other modified people around, including Hanford. But you're not transgenic, Ayden. Most of our work involved combining sequences within the genetic diversity of each species. Your DNA is 100 percent human."

"How would you know? You never confronted Gardner. He could have put almost anything in there. In me."

"Sweetheart, believe me, I've spent years studying your DNA—" Cate stopped as if slapped and shrank from the furious glare Ayden cast in her direction.

"Oh yes, of course you have. I'm not allowed to be anyone else's lab rat, but I can be yours." Ayden stood.

"Ayden, no!" Marc also stood. "Your mother was just trying to work out what Gardner had done, to make sure you'd be okay. You know that!"

Ayden didn't respond.

"Ayden, I swear to you, we had no idea he was experimenting with people. Please tell me you believe me."

Silence.

"In any case, Ayden, it was all my fault. I talked your mother into selection. If I hadn't done that, Gardner wouldn't have had the opportunity to manipulate your DNA."

"I have to be on my own for a while." He headed for his room.

"Ayden, wait."

The door crashed shut behind him. He slammed his backpack onto the bed, began fishing around in the closet for the rest of his gear.

It was unbelievable. Back at the turn of the century, molecular neuroscience was in its infancy!

He needed time to think, space to breathe.

And when they found out that Gardner had been applying their analyses to people, to their son, they still didn't have the guts to alert the authorities, to stop the maniac from experimenting with human beings.

His entire life they'd kept it from him, lied to him.

Ayden ripped his sleeping bag from its airing hanger, which sprang back, flicking across the small room to clatter against the window opposite.

"You okay in there? Ayden?" Marc's voice came muffled through the door.

"I'm fine."

"Okay, son."

[Rental car order, ex-urban usage, off-road capability, extended range, immediate availability, delivery this address.]

[Found. In price order. Daewoo super-sport-utility, hybrid electric/hydrogen fuel-cell, three-door, five-seat. Cost including insurance—]

[Accept. Delivery now.]

[Estimated delivery in twenty minutes.]

Most publicly held wilderness in the United States was zealously guarded by rangers fighting a losing battle against homeless squatters and poachers, but Ayden was a registered field researcher with extensive permits. He'd drive and hike into the North Cascades National Park, maybe do a climb.

Mainly, he'd sit quietly in the pristine forest and think. It was one of his favorite places in the world. The mountains were stunningly beautiful, and their sheer size and grandeur usually reminded him how insignificant most of the worries of humanity were.

When Ayden emerged from his room, backpack fully laden, his parents rushed to him.

"Ayden?" Cate said. "No matter what, you have to believe that we love you with all our hearts. You have to believe that. Every decision we've made—even some very bad ones—we were trying to do what we thought was best for you."

Ayden headed for the front door.

"C'mon, Ayden, say something," Marc said.

He spun around. His father flinched. "What did you have in mind?"

"I don't know, tell us we're idiots, tell us…you hate us. Anything."

"Why? Would that make you feel better?"

"Just tell us you're okay then," Cate said. "Please?"

"I'm going now."

"Where are you going?"

He walked away.

"Please be careful," he heard her small voice behind him say. "Someone might still be watching."

Sunday, May 8th

Ayden sat on the edge of a precipitous ridge. Looking down, he could see his tent nestled in the heather by the side of a small mountain lake where there had once been a glacier.

He stared, unseeing, through the achingly vibrant beauty of the wildflowers, ignoring the majesty of the mountain peaks soaring overhead and the crisp, clean air that dived into his lungs with every breath.

He was a dangerous and premature experiment. He literally embodied the worst of science.

His gut churned with fury. He could actually feel the anger surging through his veins. It was the stress hormone noradrenalin, of course, but it felt like ice and fire at the same time. Or acid.

How was he supposed to live his life?

Could he continue to challenge the hasty release of novel modified organisms without sterilizing himself?

He shivered. He'd been sitting on the icy rock for too long. And—stupidly—he hadn't eaten for over twenty-four hours. He should go back down to the tent and get something to eat. He rose and hiked along the ridgeline, heading higher.

[Wake up,] he subvocalized.

[Awake. Warning: bandwidth is very low. Expect delays,] his earpiece said.

Ayden looked around. He was lucky to get any coverage at all. He wasn't wearing his specs anyway. [Cut visual data, text only.]

[You have four new voice messages and sixteen unread text messages.]

[Who are the voice messages from?]

[Two from Cate Walker, one from Marc Walker, one from Tess James.]

[Ignore all. Full privacy mode.]

[Full privacy mode confirmed.]

[Wikiprofile—Andrew Gardner.]

[Andrew Robert Gardner. February 24th, 1959 to April 5th, 2017. American scientist and entrepreneur, he was a doctor of science in microbiology and the founder, Chairman and CEO of Sanderson Health which included the Sanderson Reproductive Health Research Hospital and the Sanderson Molecular Biology Institute, in addition to chains of reproductive health clinics in the United States, Europe, Africa, and India. Gardner was a vigorous advocate—]

[Stop. Contents.]

[Early life and education | Sanderson Health | African philanthropy | India | Personal—]

[Stop. Early life.]

[Gardner was born in Oakland, California. His parents were Harry and Celia Gardner. Harry Gardner owned a highly profitable scrap metal recycling business. Celia was a nurse at the Oakland children's hospital. Andrew Gardner attended Oakland Technical—]

[Stop. Skip education.]

[Both parents died in a car accident in 1985. Gardner inherited the family home, the business, and a $23 million stock portfolio. Gardner sold the business and began funding biomedical research, including but not limited to his own. This included: IVF research and clinics, retroviruses including HIV, cancer, highly conserved regions of non-coding DNA, and the regulatory roles of RNA. He participated in revenues from gene sequencing and synthesis technologies, which were to become the standard worldwide, and later from advances in the fights against AIDs, childhood leukemia, and prostate cancer. His subsidizing of reproductive health clinics in Africa and India brought him great praise as a philanthropist and—]

[Stop.]

Ayden didn't want to hear another word about Andrew Gardner, philanthropist. Yeah. Maybe. But all those IVF and reproductive health clinics would also have given him access to untold numbers of embryos marked for destruction or listed as non-viable.

Embryos for his parents to analyze, for Gardner to experiment with.

Ayden hiked on in silence.

Even as a child, Ayden had been able to intellectualize his problems, to derive solutions with an almost mathematical rigor.

Always.

But not now.

Worse, he knew he was being melodramatic. Unlike many in his generation, Ayden had always had a roof over his head. He'd never needed to worry about where the next meal was coming from and he'd received an excellent education. But, inexcusably for someone who had so many advantages in life, he was indulging in self-pity.

The knowledge didn't release him.

He hiked hard and fast.

Monday, May 9th

Ayden forced himself to eat something. Today would be dangerous enough without his body being weak from lack of food.

It was just light and Ayden's hands hurt with the cold as he laid out his climbing harness and attached the associated plethora of safety gear, his ropes and carabineers, and nuts and quickdraws.

He set out for Dog's Breakfast. The last time Ayden climbed Dog's, he had been led by more experienced climbers who had set the gear, belayed him, and advised him on the more difficult sections. Today, Ayden would climb it solo.

As he walked, he tried to focus on the climb ahead, to visualize the route, but his mind continued to betray him, kept slipping back to his parents, what they'd done, what he was.

It was mid-morning when Ayden arrived at the bottom of the climb and looked up. Max Abbot, who'd named the climb and introduced Ayden to it, had described it as only of medium technical difficulty "but it's psychologically challenging".

In other words, it was scary. Even the base of Dog's was elevated and exposed.

Really he shouldn't be doing this. If he fell, the gear he would set at intervals throughout his climb should save him from crashing to the earth, but in practice, Ayden knew the jerk of a 180-pound body falling to the end of the rope's slack was sometimes enough to wrench the gear from the rock. There'd be more gear below, but the odds of each piece failing would grow as his momentum built. If the first couple of pieces of gear failed, a solo climber's odds of survival were not great.

Well, there was nothing like a little visceral fear to focus the mind.

Ayden pushed his palm against the rock. The sun had started to warm it. He shook his head. He was stalling.

He checked his harness, chalked his fingers, and started climbing. For over an hour, his focus was purely on the placement of fingers and boots, on setting gear securely into crevices, and balancing on precarious supports.

Finally, he approached the summit and one of the most technically demanding parts of the climb.

Ayden found a good-sized crack to wedge his left foot into, securing it with a mildly painful twist.

Legs splayed wide, his right foot found a knob to paste onto.

There was another crevice just above his head. It was a good place to rest for a moment. He jammed his left hand into the crevice and unhooked a climbing nut from his harness with his right.

He pushed it into the crevice, then pulled on the nut's wire. It felt secure. He clipped on a quickdraw then clipped his rope in. Leaning his body against the rock wall, Ayden relaxed and let his right arm dangle, gave it a gentle shake, then pushed his right hand into the crevice above and gave his left arm a rest.

Time for the last pitch.

It was a short, smooth section of rock leaning past the vertical so he could no longer lean into the cliff to minimize muscle fatigue.

He allowed his body to lean away from the rock face, taking all of his weight on his left hand and right foot, his left foot only able to provide stability.

He stretched to the limits of his reach, his right hand seeking the hold he couldn't see but knew was there. The hold was a small rounded nub, one which would have to take his entire body weight as he hauled himself up onto the safety of the ledge above.

Ayden could feel his left hand slipping.

He couldn't find the hold. There was only one thing to do. He needed to back down to a point where he could rest again. Or give up.

To hell with that. He snarled, his left forearm screamed with pain, he felt around desperately for the handhold above.

Finally, his fingers located the hold above, just as his left hand gave way.

Ayden's body started moving away from the rock face.

Time slowed.

He didn't have a proper grip on the hold above.

He was inexorably falling away from the rock. Even if he were able to move his sluggish body at the accelerated rate of his perceptions, there was nothing to grasp onto, nothing to reach for.

He could only observe as he fell.

Ayden was now in empty space.

It felt as if his heart stopped, as if his blood paused in his arteries. His entire existence reduced to waiting to see what would happen.

The last piece of gear he had set was about two yards below. Allowing for some slack in the rope, that meant he had approximately five yards to fall before the rope would pull taut on the gear.

Five yards for his body to accelerate.

It took forever.

Ayden felt his harness grip him, jerking him roughly then suddenly he was in free-fall again. The gear, now above him, ripped out of the crevice, spraying his face with dirt.

Time suddenly accelerated as Ayden fell another three yards until the next piece of gear caught.

This time Ayden came to a complete halt. Immediately, he found foot and hand holds. He pressed his body into the cliff face.

And breathed.

Ayden's blood pounded in his ears. His body shook. His face flushed with heat.

He wanted to shout, laugh and cry all at the same time.

Intoxicated on adrenaline, Ayden felt his self-pitying fugue burn away. Regardless of how he had come to be, he was alive, vital, energetic, intelligent, capable. He would live, laugh, puzzle, rage, and contribute. He would *do*.

Setting extra gear, Ayden carefully made his way back down to earth, happy to leave a few nuts in the rock, tiny monuments of his respect for the mountain.

On his way back to camp, Ayden thought about his parents. He searched for compassion. They'd been foolish and weak, but their intentions had been pure, and they'd paid a hefty price.

[Voice call. Dad]

[Connecting.]

"Hi Ayden."

"Hey, Dad, is Mom okay?"

"If you're okay, your mother will be. Are you?"

"Yeah, I'm fine. Just needed to bash my head against a wall for a while."

"Good. So let me guess, you want every scrap of data we have."

"Yep."

"No problem. We haven't got much, and what we have is reconstructed, but you're welcome to it."

"You weren't able to take any with you?"

"I tried. Thought I was being clever. Hid the data your mother considered most important, but they must have gotten wind of what we were doing. I don't know, maybe they were snooping on our keystrokes or something. Maybe they were just being thorough. It was all encrypted, of course, so he couldn't have known exactly what we were trying to get out, but he could tell there was something there, so he just had the lot scrubbed clean. I took the drives to a forensic specialist, but Gardner used professionals. Multiple rewrites of junk over the erased data. Reformatting. The works."

"So what can you tell me?"

"Not online, Ayden. Come home. But first, call your mother."

He forgave them, but he could feel the kernel of anger pushed deep down inside. He'd never forget.

Tuesday, May 10th

Marc and Cate sat on the couch opposite Ayden. They were holding hands, had been since they sat down. Ayden made an effort to keep his tone soft.

"Is there anything else? Anything at all? Tell me now. No more secrets."

"There may be," Cate said, pausing for a moment, visibly steeling herself. "It's something I came across only recently. Your telomeres. They seem to be very long."

They stared at each other through a long silence. "Did any of your research focus on longevity?" Ayden asked finally.

"No, but we were only one of many teams working for Gardner."

"Any chance I'm just a natural outlier?"

"It's possible. Certainly, we haven't the faintest idea how they could have done it. It's notoriously hard to lengthen telomeres... in any animal."

"Perhaps I'm transgenic after all. Perhaps I have turtle telomeres or something. No, don't panic, I'm not going off grid again. I think I'm beyond shock at this stage. So, I'm not up on biogerontology, what does it mean for my life expectancy?"

"We have no idea. There's no applicable data," Cate said.

Marc leaned forward. "Not that the facts would stop the loonies for even a moment. Can you imagine what would happen to you if this all became public knowledge? First thing I did was look up maximum human lifespans. There's a lady in Illinois who celebrated her 127th birthday the other day. Then I looked up maximum mammal lifespans. Some tortoises and whales make it to 200 or more. That'd be the headline: 200. And if somehow you managed to avoid the whackos trying to put a bullet in your brain to buck the odds, you'd become the focus of jealous hatred for billions. They'd talk of the need to root out the uber-humans, unnaturally intelligent and long-lived, monsters created in defiance of God."

"Hmmm…you may be right."

Silence. Minutes passed while Ayden, Marc, and Cate each contemplated what might have been, what might yet occur.

Finally, Ayden spoke. "What I find really amazing, well, one of the things I find amazing, is that you two have lived with this hanging over your heads for all these years."

"We had each other. We shared the secret and our fears. I think I'd be batty by now if it weren't for your father."

"Whadaya mean, would be?" asked Marc, forcing a smile. It earned him a playful but hard slap on the shoulder. "Ouch," he said, grabbing Cate so she couldn't do it again, a movement that ended in a hug.

Ayden managed a smile but didn't join in.

● ● ●

Ayden was out on the tiny balcony again. It was overcast, of course, and a damp, musty smell pervaded, but it wasn't actually raining and the little fiber-resin chair was dry.

As far as his parents could tell, Sanderson's had confined itself to conventional research after Gardner died from a massive stroke in 2017. It appeared that the madness had stopped as quietly as it had started.

Hopefully.

But something was going on now, something that involved William Hanford and the tangled web of interests associated with

Genenco. There was almost no information about William, which left Ayden at a disadvantage in dealing with him.

And unfortunately, despite "everyone knowing about Genenco," there was very little information in the public domain beyond the superficial.

Though it was a diversified, multinational biotech business, Genenco was privately owned, not subject to the disclosure requirements of publicly listed companies. It wasn't clear who ultimately owned or controlled the company. Its share register was composed almost exclusively of trusts.

Management were not required, or even permitted, to report their deliberations to any group other than the non-executive board of directors, which was composed of representatives of the owner trusts. The extent and nature of research funded by Genenco was never disclosed. The public only ever found out about products once they were seeking regulatory approval, or commercialized.

The bulk of Genenco's public profile was extremely positive. It funded a network of family health clinics throughout the Third World. It demanded and received a great many concessions from the host governments, including those governments throwing their weight behind preventative measures such as the use of condoms, much to the anger of some churches. These measures had a demonstrable effect on sexually transmitted diseases and population growth rates.

Genenco also championed environmentalists' calls to save ecosystems, though once again their methods were frowned upon by some. Genenco favored genetic engineering and other proactive interventions to achieve results where traditional conservationists called for humans to leave key areas alone to fix themselves.

Ayden was still at a disadvantage. He still didn't know enough, but he'd have to work with what he had. He messaged William.

[Good timing, Ayden. I'm in Seattle on Friday so we can meet in person, come to the Radisson again, 3:00 p.m. Ask for me at the front desk.]

Friday, May 13th

They met in the Radisson penthouse. It was beautiful, probably the most opulent room Ayden had ever been in, but pedestrian compared with their earlier virtual meeting space. There were orchids everywhere though, and their delicious, slightly spicy fragrance added an atmosphere that no amount of dramatic virtual scenery could produce.

"You're wearing your phone," Ayden said as William waved him to a huge, deep aubergine armchair and sank into its twin opposite.

William's aide had required Ayden to shut down his phone, had even run a scan to confirm the cessation of artificial electromagnetic activity around Ayden's body.

"It's not impossible to make your phone secure, just expensive, and restrictive if you're the sort who likes playing with the latest apps. One moment." He took off his specs and earpiece and put them on a little side table. "Off. An even playing field. So how much do you know?"

"I know my DNA was manipulated with the aim of increasing my intelligence and possibly lifespan, and you're probably the same. You were right to make me hear it from my parents, by the way."

"You would have gone back to them for confirmation anyway. Better for them to tell you in their own words."

"And you didn't want my anger focused on you. Still, it was better for me to hear it from them. Thank you."

"In any strict sense, we're not related, Ayden. But in some ways, we're closer to each other than to anyone else. The tiny sequences of DNA we share are the products of our parents' minds rather than simply their bodies, and those sequences set us apart from all others."

"Which raises the question: How many of us are there?"

"That depends entirely on what you mean by 'us.' As in all genetic research, they only brought a tiny proportion of embryos to term. Nevertheless, there are several thousand people alive today with sequences engineered for longevity or rationality or both."

Ayden was silent for a few moments, his eyes unfocussed. It was horrifying and fascinating at the same time. "It didn't even occur to me to think on that sort of scale. They must have used tens of thousands of embryos."

"Hundreds of thousands, at least. They had access to a worldwide network of IVF clinics and were in the forefront of artificial embryo twinning for therapeutic stem cell research."

"Thousands of people like us." Suddenly his eyes were very much focused. They bored into William's. "I want to meet some of them. More of them, I mean."

"I didn't say that. Most interventions that were brought to term had modest or no measurable impact on mental functioning. Some resulted in high-functioning autism or personality disorders. It was only the closely controlled cases that produced outstanding results, and even then only in a small minority. Even the uterine environment had an impact. It was important for the mother to ingest the right substances, for her body to provide the right mix of fatty acids, for her to maintain low stress levels, et cetera, et cetera. And, of course, as much of the development of the human brain occurs after birth, even the young child's environment and diet had an impact."

"So how many?"

"There are a few hundred people walking around who are clearly more intelligent than average and a similar number who may live significantly longer than average, but their rationality quotients are within normal ranges. Fundamental success was much rarer."

"How many?"

"Just us."

"Two."

After sitting down, neither man had moved. Now, Ayden rose and went to the window. He stared out over Seattle's cityscape.

"And what, hundreds at least, possibly thousands…mentally damaged," he said. "It's unbelievable."

"It sounds bad, yes, but only around 25 percent of those born have had significant mental disorders, compared with nearly 20 percent in the general population. I think they managed the downside admirably. And they did finally achieve success."

Ayden shook his head and turned to look at William. "You used the term 'fundamental.' How do you justify that? Surely, we're just at one end of the distribution curve of those few hundred."

"We're much more than that—we're complete outliers. And in one critical capacity, we're different from anyone ever measured, almost certainly from anyone who has ever lived. The biggest measurable difference is in our working memory capacity. Human beings can only hold three or four items of information in their working memories. All other feats of memory involve recall from long-term memory."

"Of course."

"No one has ever demonstrated any evidence of a working memory capacity beyond six or seven elements."

"Okay."

"Except you and me."

"I've never been involved in any specific working memory testing."

"You did an aptitude test when you went to college. You may recall some of the details. It was an adaptive test. Yours was special, based on my results."

"I remember being impressed with the test, with how tough the questions were, toward the end."

"Your results match mine, which I've tested extensively."

"And?"

"Twenty to thirty, depending on the testing regime."

"Nonsense. We must simply be better at shuffling items to and from long-term memory."

"Examine the tests and results yourself. I'll send you the data when we're back online."

"I will."

"For the purposes of this conversation, assume that you'll accept my results."

"Okay. A minute ago you said 'rationality quotients.' What do you mean?"

"Gardner looked at combining the old notion of IQ, or intelligence quotient, with EQ, or emotional quotient, which was supposed to measure emotional stability. I've refined his ideas a bit. Acting rationally requires both practical intelligence and emotional intelligence. Your RQ is your capacity and propensity to act rationally, particularly in complex and stressful situations."

Ayden returned to the chair opposite William and sat.

"I'm still missing something. Okay, so we have outstanding working memories and intelligence measures, that would be very impressive, but I'm not sure it would qualify as fundamental. There have been lots of outstanding people in history. What piece of the puzzle are you withholding?"

William smiled for the first time. "The key. Philosophers have struggled to define consciousness for centuries and still don't have final answers, but science has proven that when people are faced with complex, uncertain situations, they make better decisions when they 'sleep on it' or 'think about something else,' or 'follow their instincts,' euphemisms for processing information and making decisions subconsciously. Like the rest of the animal kingdom. The much vaunted human consciousness is largely relegated to being a language-based internal narrative that makes sense of the external world and of our subconscious responses to it." William paused, giving Ayden the opportunity to contemplate the implications of what he was saying. It didn't take long.

"You're hypothesizing that working memory is the bottleneck which limits our capacity to reason consciously. That it's a constraint on our rationality."

"Correct. Homo sapiens' capacity to out-think animals is ultimately limited by a four item working memory, but our capacity, yours and mine, to out-think the rest of humanity is provided by a twenty to thirty item working memory. I'd say that constitutes a fundamental difference."

"But if it was so successful, why didn't they keep going?"

At that moment, there was a timid knock on the door.

"Excuse me, he's not used to me being off-grid without warning," William said, then he called loudly. "Yes?"

The door opened part of the way, and William's aide stuck his head around it. "Sir?"

"It's all right. I've just turned it off. Water for me. Ayden?"

"Orange juice, please."

The aide's head disappeared.

"Their own limitations sabotaged them," William said when the door fully closed. I was the first real success. You were one of fourteen very closely controlled attempts to replicate my outcome. Only you showed early signs of it having worked. There was still something missing, something your parents and mine supplied naturally, a subtle variance in how our genomes interacted with the introduced sequences. Based on their analysis of us, they thought they'd identified a missing regulatory sequence. They engineered it into the next batch."

"And?"

"Psychosis."

"You said they'd caused mental disorders before. Why did they let it stop them this time?"

"They'd become confident and were working with their own material."

"They made their own children psychotic."

"The team's reaction was, of course, irrational. Despite their previous successes, the failures in their own children were intolerable. Rather than try again, reanalyzing what happened, the team collapsed. They wrote off the whole project as a failure. All except for Gardner. He never gave up. He viewed the failures as valuable data and redoubled his efforts to understand the limits they had bumped into.

"It took him nearly five years to convince himself he understood what had gone wrong. I get the impression from his notes that he thought it would have been quicker if your parents had still been on-board. Together they did brilliant work, very subtle. He was recruiting another intervention team when he died."

"Were your parents like mine, or did they know what was really going on?"

"My biological mother was part of the original team."

"And your father…"

"… was Andrew Gardner. They weren't a couple. Gardner was single, and my mother's husband had no idea what she was doing. They used a surrogate to bring me to term."

"And you inherited Gardner's wealth."

"His personal wealth, yes. But let's be clear on this—I don't own or control Genenco."

"Who does?"

"Essentially, it's owned, in unequal share, by the members of about a dozen extended families, but the exact holdings are almost impossible to determine because of a web of interposed trusts and companies. I own a relatively small chunk of a behemoth."

"Who controls it? Do you?"

"That's like asking who controls the government. No one person does, not even the president. Genenco's big, Ayden…it's the second-largest unlisted private company in the world. I'm a player though, a trustee for one of the holding trusts and, as well as the money, I inherited my father's notes and they've been invaluable. Information can be more powerful than title deeds."

"So that's the life you've chosen, the corporate animal."

"It's a means to an end, one which consumes about 20 percent of my attention. More importantly, I'm involved in some of the most fascinating scientific advances on the planet."

There was another soft tap on the door, and William's aide brought their drinks in, then left without a word.

The pause stretched. They studied each other.

"Are you continuing Gardner's experiments?" Ayden said. "Trying to create a better human?"

"No. His team, including your parents, created us. Now it's up to us to decide how best to apply our abilities."

"What about illegal experiments with other organisms, like mosquitoes, for example?"

"Is that really what you came here to do? To hurl allegations about insects at me? Let's talk about the things our parents did. Let's think through the implications of who we are, because I think we can be powerful allies. I think we can be a tremendously powerful force nudging this planet in rational directions."

"How do you plan on going about that?"

"With science, of course. Half a century ago, Robert Solow won a Nobel Prize for proving that 90 percent of real economic growth comes from technological change, a fact that's been used to help justify almost every R&D budget and tax concession since. So, nudge science in the right directions, and you can affect the decisions that nations and multinationals make."

"It's as simple as that."

"In the very long term, yes. In practice, I'm not sure we have the luxury of that much time."

"Don't you think it smacks a little of megalomania that you think you can save the world?"

William regarded Ayden silently for a while. Finally, he spoke.

"All your life Ayden, you have underestimated your capabilities. Don't make that mistake with me. Unlike you, I haven't been blinkered by parental obfuscation. I've been learning what we're capable of. Effectively, I've been experimenting with my ability to maximize my capacity. I've placed myself at the hub of an extensive information-gathering machine."

William leaned forward, fixing Ayden with his eyes.

"No one else, save you Ayden, could possibly absorb and use the amount of information I do, and I have very substantial resources at my disposal—both tangible and intangible. Think about that, Ayden. An intelligent and capable person can only achieve so much on his own, but give that person influence over the resources of a multinational corporation and see what they can achieve."

Ayden barked a laugh. "That's your response to being accused of megalomania?"

"You'll have to make up your own mind about that. Do you see me parading around like a power-addicted, alpha male, homo sap? Trophy women hanging off me, with their gene-pumped breasts? Palatial yachts and million-dollar cars? Surrounding myself with 'yes-men' to reassure me of my stature?"

"No," Ayden said.

"Believe me, I've got the money. But I don't care about that rubbish any more than you do. And now you know the truth about yourself, you can cast aside your prejudices against genetic engineering and assess each case on its merits."

"I'd like to think that's what I've always done. Maybe I'm just more risk averse than you."

"C'mon, Ayden, we've been artificially manipulating the evolution of plants and animals for thousands of years. Genetic engineering's just the 'power tool' equivalent of selective breeding. And just because people have managed the risks badly in the past, doesn't mean we can't do it properly. And when I say 'we,' Ayden, I'm not referring to humanity. I'm referring to you and me."

"Why?"

"Because Gardner's experiments worked! We're more capable of assessing large amounts of information and more able to respond rationally to the complexities and uncertainties of the world around us.

"Work with me, Ayden. Embrace what you are. Use it. Help me to help mankind. Look at the mess they're making of the world. They're killing each other as individuals, going to war as nations, consuming without thought or limit—putting unbearable loads on the planet's resources, destroying environments, wiping out irreplaceable species, and playing havoc with the world's climate."

"True." Ayden looked down and shook his head. "Too true."

"And to what effect? Even at a huge cost to the planet, they're not making themselves content. Once they're wealthy enough to be safe and healthy, further wealth has no impact on happiness. That was scientifically established decades ago, but it hasn't stopped them. They're helplessly driven by millions of years of evolution to crave status through the display of wealth, all because primates maximize their reproductive success through social status."

Ayden looked up. "Are you claiming that you and I don't share those instincts?"

"I'm saying we don't let them rule us. The problem is, homo sapiens' decision making has a massive Achilles' heel. They're okay if the situation is complex, as long it's not emotionally loaded, or if the situation is emotionally loaded but simple, but they make very poor decisions in complex situations charged with emotions."

"Not all of them."

"No, but it's their collective actions that'll destroy them and wreck the biosphere. We have a responsibility to help them overcome their shortcomings. And that's the other thing, Ayden. Time frames."

Ayden felt a shiver run through him. He put down his glass. "What do you mean?"

"What do you think I mean?"

"You're talking about our telomeres."

"Yes. We may well live much longer than them, and we're rational enough to plan for our long lives, to think about the implications, to make the world saner. Imagine one change, one rational decision made worldwide—imagine that every nation on the planet ceased all military spending—and allocated that money to addressing the long-term issues the world faces."

"A lovely fantasy, I agree, but not achievable in our lifetimes. Possibly not ever."

"Not under current conditions, perhaps, but under the right conditions? Given enough time? Look at the impact of international trade between former enemies. Of the communications revolution. Even now, developed countries only go to war with those nations that are of little value as trading partners. Conflict between major nations has become uneconomic. Look at what the grid is doing to communication between individuals. It's robbing leaders of the capacity to lie to their people with impunity, allowing everyone to see how everyone else lives. Where the grid dominates, dictatorships fade away. Wait a while. Can't you see the idea of war becoming a complete anachronism?"

"That depends. How much time are we're talking about?" Ayden asked, leaning forward.

"I suspected that you only knew half the story," William said. He, too, leaned forward, fractionally.

"I know our telomeres are longer than usual, but my parents had no idea what that would mean in terms of lifespan."

"Our cells produce telomerase, which largely stops telomere shortening."

"Don't cancer cells achieve immortality the same way?"

"Yes, but primate testing convinced my parents that our modifications didn't increase cancer risk, besides which we have multiple copies of an aggressive version of the tumor-suppressing TP53 gene."

"How long?"

"We can't know for sure, of course. We're effectively initial experiments. I can only make broad estimates."

"Your access to Gardner's data means you can make a better guess than anyone else. How long would you speculate?"

"In addition, our bodies produce oligomeric procyanidins, which provide substantial protection to human blood-vessel cells, protection against cardiovascular disease."

Ayden waited.

"Based on the chimpanzees, a hundred and fifty years easily, barring accidents, of course, and with an appropriate lifestyle," William said. "Two hundred maybe. And with sufficient medical advances over the next hundred years or so? Another century on top? I suspect mental deterioration's the main threat, but I wouldn't be hurrying to set any upper expectations. Some trees live for thousands of years."

"Show me the data."

"One step at a time, Ayden. It's not much better than an educated guess in any case, but it does make you think differently, doesn't it."

They stared at each other, unmoving, eyes locked for an age.

Finally, Ayden broke the silence. "Yes."

"About the mess the world will likely be in a couple of hundred years from now."

"Yes."

"About what we could achieve."

Ayden contemplated William's words. A time frame measured possibly in centuries certainly would allow someone to contemplate ambitions that would otherwise be fantasy.

"The question becomes what achievements? Whose goals? Yours? Mine?"

"Oh, come on, Ayden. We're not chest-beating alpha males, stomping on prone betas, fighting each other for dominance. If we're as rational as I think we are, it'll become an easy alliance of equals, seeking ultimately to make the world a better place to live for everyone, deciding on strategies rationally."

"But, initially, you don't trust me enough to tell me all about your operations and, no offense mate, but I don't trust you enough to support you without reservation, so we recognize that and…?"

"We take one step at a time. We assist each other as we see fit and feel comfortable. Over time, we'll come to understand and trust each other implicitly. Initially, my proposal to you is this— your bees. I want it solved as much as you do, probably more given that Genenco is implicated. Our top priority must be to solve the bee pollination problem, and I can't think of anyone better to head that effort than you. So how about this: I use my influence to have your role expanded to include intervention. I give you access to additional information and resources. You find a solution. We fix this thing."

"But I shouldn't expect you to tell me what happened with the mosquitoes or how the bacteria spread to bees."

"How much would be achieved by dragging your parents over the coals for their actions two decades ago? How much would be achieved by associating this bee issue with old Genenco research projects?"

"Knowing what happened might assist in determining the best intervention."

William sighed. "I've looked into it. I don't know how the bacteria transferred to bees. I can only assume natural horizontal transfer. Of course, when the modified bacteria transferred to blood-sucking mosquitoes, MTC terminated their plans to commercialize Rhino Mozzies. It was a screw-up, which the people responsible for tried to sweep under the rug. They've since been fired, their careers haven't recovered. What more should we do to them?"

Ayden didn't answer.

"I'd suggest engineering a bacteriophage, a virus specifically designed to prey on the modified bacteria, and seed it amongst the bees."

"Another GMO sent out to contain a previous GMO which broke containment."

"Yes, Ayden, but you could do it properly, even if most others couldn't."

"Maybe."

"And if there was a problem with your GMO, then I'm sure you would address it, perhaps having built in some 'hooks' for attacking it, if required. There can be no progress without risk, Ayden."

"And, if I can find a non-GM solution?"

"Then you earn my admiration, because I doubt that I could."

"Let me think about it."

"Of course. Take as long as you like."

Monday, May 16th

Three days later, Ayden was still deliberating. His earpiece chimed. [Video call from Linsey Carr,] it said.

[Accept.] "Hi, Lin."

She looked worried.

"Hi, Ayden. You're in Seattle then?"

Ayden had authorized team members to have access to his general location, "general" defined as within a five-mile radius.

"At the condo. What's up?"

"It's George, Ayden. He's wanting ye here. Right now."

"I'll give him a call."

"Nooo, Ayden." [You really don't want to be calling George this time. He took a call from the EPA this morning, has hardly said a word since.] "He specifically asked for you to come to the office." [My advice would be get your butt in here, like ten minutes ago.]

With a flick of a finger, Ayden checked the condo's garage. There were three cars available with plenty of charge, one of which was free of bookings until early evening.

"I'll be there in twenty minutes."

• • •

George gestured for Ayden to take a seat on the other side of his broad antique desk, which contrasted strongly with, but somehow complemented, the rest of his unrelentingly eco-modern private office with its compressed cellulous boards and translucent panels.

"Specs off, please. I want your full attention," George Reyes said.

Ayden complied without a word, but left his earpiece in. He found himself looking upwards at George just slightly. No doubt the visitors' chairs were deliberately low. George knew every corporate gamesmanship trick in the book.

"It seems I'm not in any danger of boredom with you around, Ayden."

"Sorry?"

"This morning I received an offer from the EPA to extend our role to intervention."

"That's great news."

"...specifically nominating you to lead the intervention."

"Ah." Hands hidden under the desk, Ayden activated his gesture cuffs. [Message to William Hanford. You said I should take my time,] he subvocalized.

"Yes, 'ah.' What's going on this time, Ayden? You appear to make weighty friends as quickly as you make weighty enemies. Generally in a situation like this, I'd be letting the person in your shoes know—in no uncertain terms—that I don't take kindly to being side-stepped. In your case, I'm going to hear you out first. I can't believe my character judgment is that poor."

"Thanks, Dr. Reyes. I certainly haven't intentionally circumvented your authority."

[Text message. Response from William Hanford reads: "The opportunity arose so I took it. Feel free to say no."]

"But," George Reyes said.

"But...well, does the name Hanford mean anything to you? William Hanford?"

"Rings a bell...no perhaps it doesn't. Should it?"

"On the board of Genenco. Keeps a low profile but very influential. I'm only just starting to appreciate how influential."

"Genenco? I can't wait to hear the rest of this"

"There's some...history. Between us. Between our families. I can't go into the details, but in short, he offered me his 'assistance' in expanding the role to intervention."

"I'm going to need more than that."

"After the MTC mess, William contacted me and we met... and he made me the offer."

"You shouldn't have met him without clearing it with me."

"I apologize."

"Do you? Hmm. And the connection with your family?"

"I'm sorry, I can't talk about that right now. I said I'd think about his offer. I figured that, once I'd thought it through, I'd speak to my parents, then you, before going back to him. I had no idea he'd just go ahead, or that he could so fast."

"Leaving aside your little family mystery for the moment, what does he get out of the deal?"

"That's the dilemma I've been struggling with. It's all a bit vague, but Genenco was involved in the bacteria transferring to common, bloodsucking mosquitoes—accidentally, he claims. So they're at least indirectly responsible for the bee problem."

"And he doesn't want us—you—to pursue the issue of responsibility."

"Correct."

"That's easy. Accept the role."

"It's that straightforward?"

"Of course. If the evidence is absolutely clear, it'll get picked up. If not, then Genenco will be untouchable. It's a tremendously influential organization both here and overseas. Its legal and PR machines are second to none, and many governments are beholden to them for one thing or another. You can achieve something, however, something far more important really. You can make sure the intervention is done properly and effectively."

"Be pragmatic, in other words? To hell with ethics and justice."

"I believe the concepts of ethics and justice are among the highest achievements of the human race, Ayden. They're to be revered and sought at every opportunity. But they're goals. Targets we usually fail to hit. If you want to actually achieve anything, you pick your battles. Don't lose your idealism, Ayden, but be realistic in your expectations. And patient."

"So you'd be happy for me to accept?"

George Reyes' round face creased with intensity. "Happy is not the word I'd choose here. I don't like the way this has happened, but there's something I want you to understand about me. You know, of course, the history of the U.S. finally getting serious about climate change?"

"America's worst natural disaster."

"Well, yes, in some ways. What the media dubbed the 'Year of Storms' was two years before Zelda, and it killed over 35,000 people. It wasn't just one storm though, it was a series of tropical cyclones and floods, so I'll blame that for its failure to galvanize action—rather than the fact that the victims were mainly in developing countries."

George Reyes stood and began to pace. Ayden turned his chair a little so he could follow him without craning his neck.

"I can scarcely describe how frustrating those years were. The Intergovernmental Panel on Climate Change had just delivered its fifth assessment—the world was falling dangerously behind schedule. The Asian economic boom driven by China and India was based on fossil fuels and greenhouse gases continued to rise. The assessment warned of imminent risk of tipping points being reached, assuming they hadn't been already.

"Despite strong rhetoric though, the Copenhagen deadlock persisted. China refused to handicap its growth while its per capita emissions were still a quarter of the U.S. level, and the U.S. wasn't about to make itself even less competitive with China, or to take on its own domestic special interest groups.

"I think the cracks were forming though. The Gulf of Mexico oil spill in 2010 put a serious dent in the oil lobby, and the Year of Storms in 2013 convinced a lot of people that all that extra energy we were pouring into the atmosphere had to go somewhere. But you're right, it was Hurricane Zelda that really made people in the States sit up and pay attention."

"Together with the incompetents who didn't get the evacuation started early enough."

"Well, there'd been a few false alarms and people were becoming complacent and the damned thing changed direction unexpectedly."

"Surely, that's no excuse."

"No, probably not, but it's a major part of the reason. The devil is inevitably in the detail in these matters."

Ayden knew the basic facts, but he'd been three at the time. Reyes had been a climatologist. "So what happened?"

"It started the previous Thursday when a powerful but unremarkable storm brewed off the west coast of Africa. By Saturday, the U.S. Weather Bureau had issued a hurricane warning for Orlando and Daytona Beach, expecting a category two, or possibly three, storm to hit on Wednesday. "But on Tuesday afternoon, the storm turned north east toward Bermuda and abnormally warm water fed the damned thing," Reyes said. "Winds increased to a category five 170 miles per hour, then around midnight it changed direction again. Contradicting all the modeling, it turned due north and picked up forward speed. By 1:00 a.m., the hurricane warning was changed to cover Delaware, New Jersey, and New York, but that didn't give them much time to wake people up and evacuate.

"The eye of the hurricane passed a few miles west of Manhattan at around 10:00 a.m. on Wednesday morning. Wind speeds had abated somewhat as the storm passed over cooler waters, downgrading it to a category 4 but speeds of 150 miles per hour were recorded as it made landfall. Of course, damage from the wind itself was nothing compared to the storm surge."

George Reyes stopped pacing. He paused, shaking his head slightly at the memory.

"Please go on," Ayden said.

"Well…the surge coincided with a high tide, reached a height of twenty feet. A mountain of water flooded into New York.

"The financial district in lower Manhattan was submerged, as was Brooklyn, Long Island City, Queens, and JFK Airport. Manhattan's subway system flooded, as did the Lincoln tunnel."

George stopped again. He looked through Ayden.

"I lost two friends. And a second cousin. The death toll was eventually tallied at 14,841. And it wasn't New Orleans this time, it was the Big Apple. In a way, it was like 2001 all over again. From the moment Al Gore's supporters talked him into another tilt at the presidency, the result was a foregone conclusion, and when he made his, 'It's my turn to take us to war' speech, he wasn't only promising the action we demanded. Gore's 'war on climate change' offered us a morally unquestionable war. And one that wouldn't involve body bags."

George took his seat again.

"The economy was still very shaky, but Obama's increases in science funding helped lay the cornerstones for addressing CO_2 emissions, so Gore was handed both a clear mandate and the means to achieve it."

"But the link between climate change and Zelda couldn't have been strong, not back in 2015. Sea levels had hardly budged. Okay, warmer ocean temperatures may have been a contributing factor, but that storm could have happened any time, without climate change."

"Sure. Nearly did, back in 1938. And the Galveston hurricane in 1900 killed proportionally more people. But what Zelda did was provide an emotional impetus to respond to the facts which had been staring us in the face for decades."

"Whether or not climate change caused Zelda, it would certainly make another Zelda more likely?" Ayden said.

"Exactly. And the world was going to make changes with or without us. Behind closed doors, business knew that if we wanted a chance to compete in the new energy markets, we had to get serious. The economics was even clearer than the climate science. Zelda was just the tiny straw that broke the camel's back. At the time, I was working on the impact of additional atmospheric water vapor caused by warming on cloud formation.

"Anyway, after Zelda, my career went ballistic. Soon the administration was asking me to look at anything and everything. By the end of the twenties, it'd become clear we needed to do more, to be proactive. I was pushing a massive international biochar funding project. I figured we could sequester carbon and address the worldwide deterioration in soils at the same time. It'd even help with the nitrogen cycle problem through reduced soil emissions and fertilizer use, but then the religious right finally put the Republicans back in power with Riley in twenty-nine."

"And he backed cloud seeding."

"And I got sidelined."

"Is cloud seeding the wrong choice?"

"I have serious reservations, but the truth is I really don't know. I guess we're going to find out. In the meantime, I'd been sitting on a few boards and the speaker circuit has been good to me, but it seemed the exciting part of my life was over. I was bored stiff when your mother called me. I understand you were behind that."

"Sort of. We wanted to pitch for the role. Well, actually, I very much wanted to. Mom agreed to try, but we didn't have the track record or credibility, so she suggested we try to get you involved and piggy-back your project."

"Well, for me it was a godsend. I'm leading a scientific investigation again, and I couldn't be happier, which is why I let an inexperienced youngster like you head up field research…and it's why I'm prepared to give you so much leeway now. I *think* I believe that you didn't engineer this little coup deliberately. And I *think* I trust you enough to work with you. Certainly, I feel I owe it to your mother to try."

George leaned forward and tapped the desk softly with his index finger.

"I'll even leave aside your little family mystery for the moment, but be on notice, Ayden Walker, I'm watching you very closely, hoping I don't need to change my mind about you."

"Thank you. I won't disappoint you."

"Don't." George Reyes sat back in his chair and smiled. "Actually, you'll be putting me in a comfortable position. If you crack it, then I share the glory. If you screw up, well then…they appointed you, not me."

"I want to get my parents' opinion."

"Do. Then let me know, and I'll accept the EPA's offer."

• • •

They sat in the little beige but leafy condominium.

"Take the role, Ayden," Marc said.

"Wait," Cate said. "Sweetheart, please be careful how close you get to Will Hanford. Don't just trust him. He's Gardner's son. He could be a saint for all we know but equally he could have followed in his father's footsteps."

"I'm giving William the benefit of doubt. It doesn't mean I trust him. And I think I should take advantage of the situation to make sure the bee intervention is done right."

"Okay," Marc said. "What can we do to help?"

"Glad you asked…"

For the next few hours, they lost themselves happily in discussing and planning Ayden's approach.

[Good decision Ayden.]

[Thanks William. Perhaps a little warning next time?]

[And risk boring you?]

Monday, May 23rd

Ayden thrust aside his personal turmoil, fiercely enjoying being immersed in science again. Although their final report would, no doubt, be subject to endless redrafting and polishing to appease their political masters, the initial bee investigation was winding up. Ayden moved fast to retain people and resources for the intervention project.

He could work from anywhere, but he'd opted to grab some of the office space that became available as George's team shrank. He wouldn't be roaming in the field for a while, and the arrangement would give his parents—and him—a little more space. He'd also managed to grab a share of Linsey's time to help manage day-to-day logistics.

The office was open plan in the 'cafeteria style' that assumed everyone worked comfortably in a desk-less environment using specs or tablets. In the central area, clusters of chairs and tables were separated by translucent, compressed cellulous panels which revealed their natural fibrous complexity when light shone though them. Around the outside, "nooks" were set into the walls, ostensibly for those wanting a more private environment. There weren't enough desks for the whole team to be physically present at once, of course, as it was assumed that some people would always be in the field or in a lab. In any case, everyone was accessible virtually as required, so those who "needed their own space" tended to claim nooks by strategically leaving possessions in them or time-sharing with colleagues, giving rise to an informal but complex set of "squatting rules." As far as Ayden was concerned, privacy and personal space were more a matter of who you invited into your virtual view than physical partitions and walls, so he tried to set a good example by sitting at a different table every day, but it didn't seem to make any difference.

Ayden decided to test himself somewhat. In a small way, he emulated the information-hub approach William had described. He set up multiple sub-teams and imposed strict discipline for their data. All their work, from rough notes to final reports, had to be hosted in the project virtuality, and extensively cross-referenced and linked to source data. He immersed himself in the data and analysis produced by dozens of people. He pushed himself as never before, seeking his limits.

He was concentrating so intently that he didn't notice Tess standing nearby.

"Knock, knock."

He flicked BEAN aside. In her work clothes, Tess was casually attractive. In a somehow-gravity-defying dress that hid much but suggested more, hair let loose on the world and subtle makeup emphasizing her naturally large eyes, she was something altogether different. She smiled invitingly and raised an eyebrow.

"Hi, Tess. Wow."

"Am I interrupting you?"

Ayden smiled and removed his specs.

"I guess you're going to ask me about your data?"

"Oh. I thought we'd forgotten about that."

"Sorry, I've been a little distracted. I have had a look though, and—"

Ayden stopped. Tess had closed her eyes and raised a hand in mock drama.

"Ayden Walker, are you aware of the current time?"

"It's nearly eight."

"Because I've met your mother and she seems like a very nice person, but she really should have taught you that all work and no play will make Ayden a very dull boy."

"Well, we can't have that. Did you have something in mind?"

"Nothing specific. A flat surface. Food. Candles optional."

"I see. Ahem. Tess, would you care to accompany me to dinner?"

"Why thank you, Ayden. I'd be delighted."

* * *

The small Vietnamese noodle house didn't have candles, but it was clean and pleasant and softly lit. The table offered their specs menu links, and they ordered quickly.

"You know workaholism is a disease too, right?" Tess said.

"I've gotta say, I'm really enjoying myself. I've never tried to juggle so much information at once before. I'm used to working pretty much alone, but now I've got four teams going flat out, and I'm trying to absorb everything they're throwing at me."

"I hear you're coming up with insights they haven't thought of. Like there're four of you, each focusing full time on a team."

"Someone needs a pay raise."

"Seriously, you're in danger of getting a reputation for genius."

"I know how this is going to sound but it feels…invigorating. Thrilling even. Like I'm really stretching my legs for the first time ever." Ayden paused, acutely aware that he might be sounding arrogant. And just plain weird. He examined her face, but she wasn't judging him, not harshly in any case, he decided with some relief. Her opinion mattered to him. *When did that happen?*

"It's like I'm feeling the cerebral equivalent of an extreme sports adrenaline rush. It's a charge."

"Which makes you what? An extreme nerd? Lucky for you, you're also cute. So, is all this impressive brain power achieving anything concrete?"

"I think so. The problem's not trivial though. We have to stop the spread of the infection, but the uninfected bees have to be allowed to continue to pollinate crops and other flora. If it were possible, and proved necessary to wipe out the infected bees, we would, but it'd take years for populations to recover. Then there's the minor concern of protecting the bees from re-infection from mosquitoes."

"So what are you going to do?" Tess asked as a waiter poured her a glass of rice wine. She thanked him with a quick smile. He then poured Ayden's. "Well, loathe as I am to admit it, we may need to use some sort of biological control."

"Why loathe?"

Ayden smiled. "Well, you know I grew up in Australia. I remember the cane toads. They were a 'bloody disaster.'" He stretched the phrase into an exaggerated Australian drawl. "They excrete poison onto the surface of their skin. The reptiles and mammals that prey on small amphibians didn't, of course, have time to evolve defenses so many of them were poisoned to the verge of extinction. At the same time, amphibian habitats were overrun, forcing a number of native amphibians to the wall as well." He took a sip from his glass. "Hmmm, this isn't bad."

"It's sweet, isn't it?" She paused. "I think I've heard of cane toads, vaguely. They were deliberately introduced?"

"Originally from South America, they were used across the Pacific to control agricultural pests, and about a hundred years ago they took some toads from Hawaii to Queensland to deal with beetles infesting their sugar cane plantations. By the time my family lived in Australia, they had spread across Queensland into the Northern Territory. Now they're all over the top half of the continent causing a major loss of biodiversity."

"Not a shining example then."

"Not one I'm keen to emulate. I need something really targeted to Ps Novo. A successful enough organism to spread through the bee population, but with a serious weakness to stop it from spreading uncontrollably outside the target population."

"And there isn't one?"

"Doesn't look like it." Ayden paused. He could hardly believe what he was about to say. "We may have to engineer a GMO to do the job."

He pretended to ignore her guffaw and thanked the waiter as two steaming bowls of noodles arrived.

"What? Mr. BioWatch anti-GM himself? Cooking up a GM to release on the unsuspecting world?"

She tipped her head back and belly laughed. It was a delightful sound. Heads turned. Fellow diners smiled.

"Oh, I'm sorry, Ayden, but the irony is just beautiful. Your friends are going to have a field day!"

Ayden hadn't been against all applications of GMOs in all circumstances—though it was very much an emotionally conflicted subject now. Where there were no less risky alternatives, GMOs were justified. The rider, however, was that extraordinary care was required to ensure that unexpected consequences were avoided.

But to defend himself would just hand Tess more ammunition to roast him with. "Yeah, yeah. Laugh it up, girl. Just watch you don't choke on your tofu."

After a while, she relented. "So am I kidding myself with this capacitation thing, or do you think I've found a real medical puzzle?"

"Ha! You're wondering if there's a paper in it, aren't you? Possibly worth switching specialties for? Get you started on the academic publishing track, maybe even a research grant?"

"Maybe," she said with an impish smile. "It's not easy for a scientist to get started these days."

"Well, you're definitely onto something. Infertility is accelerating in your data set, no question. You need to source much wider data though. Your population is too small and too narrow."

"More easily said than done."

"Establishing the linkage with capacitation is the tricky bit, of course, but you may be able to do it indirectly."

"How?"

"Your notes pointed to the link between sperm antibody levels and capacitation problems."

"Yes. The correlation's pretty good as far as the data goes."

"Okay, well SciVirt covers a very broad range of scientific inquiry, and I've done some asking around. Apparently, most clinics and sperm donor banks do keep sperm antibody data."

The Science Virtuality strictly limited its membership to those holding a Ph.D. or higher qualification in a science discipline, so it wasn't an inquiry Tess could have made on her own.

"That's good news."

"Yeah. The bad news is I contacted both of the major players in IVF in the United States, only to be politely invited to go jump. The raw client data is privileged, confidential by law, and they don't have the surplus resources to conduct the data analysis we require. In any case, the results would likely be commercially sensitive, blah, blah, blah. The responses were so similar they might have been drafted by the same 'communications' manager."

Both companies had links to Genenco, but Ayden was reluctant to ask William for more assistance.

"I'm glad it wasn't just me."

"But I may have good news. One of the European IVF companies seems much more helpful."

"Fantastic!"

"Hold on. They won't provide grid access to the raw data because, of course, if we can see it, we can copy it, and they won't do the analysis themselves. But, if you are willing to go to Prague, where Allied Healthcare's head office is, they may let you access the data from a locked-down terminal within their offices."

"Prague. Okay."

"You'd only be allowed to take away summary data, which you'd have to check with Allied Healthcare, and only publish with their permission. Think you can get yourself there?"

"Somehow… yes! My father's been pestering me to take a holiday for ages. Keeps offering me his United Airships frequent flyer miles. Not sure if I can do it over the summer break though, I've promised to spend time with my mom. I could get over there in autumn for sure though. Cool! Thanks, Ayden."

"My pleasure. They're the second-largest IVF operator in Europe. Two hundred and something clinics in seven countries. And they keep antibody data. With any luck, it'll be enough to confirm or allay your suspicions."

All in all, it was a lovely evening.

W.R.R. MUNRO

7:
JUNE 2033

A man stood looking out a bedroom window, wearing striped pajamas and holding a young child in his arms, a boy.

"What is it Dad?" the boy said, his hushed voice full of wonder.

The red-brown wall stretched from horizon to horizon and rushed toward them like a mountain range-sized tsunami.

"Nothing to worry about Ricky," the man said. "It's just dust."

Suddenly it was upon them, and the world glowed a hellish red.

"I'm scared Daddy," the boy said.

"It's okay," the man said. He swung the boy around to face him rather than the eerie glow of the red sky. "Guess what? Good news! We all get to stay home and play today!"

SARAH'S DIARY, "ANOTHER DUST STORM IN VIRGINIA. THE MANASSAS NANOELECTRONICS PLANT GOES INTO LOCK-DOWN FOR THE THIRD TIME THIS YEAR," SARAH.CASE.ID

Saturday, June 4th

[Hey Ayden. Guess what? Dad said he'd spring for a cabin! So come September, I'll be sipping from my complimentary glass of champagne in the first-class lounge of a Boeing 997 airship. It gets better though. He insisted I take a real break, so before I hop on a train to Prague, I'll be spending a week in the Bordeaux region of France, in a beautiful, little, seaside town called La Rochelle. They still have oyster farms there, on the edge of the Atlantic Ocean, and apparently they're safe to eat. Dad says I just have to try some and if I go direct to the farmer it'll only cost one limb instead of an arm and a leg. Me, I figure the same approach should also work for the local Pinot. This is one field trip you really should be on.]

The truth was, Ayden would love to join her. Tess was a delight. Intelligent, passionate, beautiful, and clearly her interest in Ayden extended beyond her fertility puzzle. But how could he let himself become involved, knowing what he was? How could he keep it from her? It would be a lie only by omission, perhaps, but a massive deceit nonetheless. The answer was simple—he could be her friend but he couldn't let it go any further. Not unless he was prepared to tell her everything.

Monday, June 20th

[We're ready for you, Ayden,] Linsey's voice said in his earpiece and an icon appeared by his shoulder.

He flicked out of BEAN and stood. In the center of the big, open plan space, George Reyes and Linsey Carr were waiting for him. The others would join them virtually. He glanced around as he walked over to George and Linsey. As usual, the office was quiet and looked as though people were ignoring each other and nothing much was being achieved, but as Ayden flicked back into BEAN's office view, the room transformed. The ceiling was replaced by a limitless sky full of 3D icons, which clustered over groups of people. Gentle flashes of light descended from the icons as people accessed files, and flashed back up as information was posted to the collaboration space. Occasionally, icons flashed from one group to another. As Ayden focused on a particular group, he was able to hear their subvocal conversation.

He sat down and flicked the conference icon, and suddenly he and his team were in natural leather-colored armchairs around a low stone table, surrounded by trees with tall, pale trunks and high canopies. In the distance, birds called faintly to one another. Though the sky was an intense blue, mist carpeted the ground and played a slow, silent game among the trees and around their chairs. The mist was beautiful, and its slow movements enhanced the tranquility of the scene but it also had a more prosaic purpose. Soft wisps helped BEAN to render a coherent scene by artfully obscuring clashing edges or angles.

Professor Sherman, whose team was examining the bacteria's transmission pathways and lifecycle on its bee host, was particularly wrapped in fog. Ayden's mother sat to his right. Her team was looking for genetic exploits. She was beaming at Ayden, clearly she relished watching her son take charge. Beside her sat Robyn Cassidy, a marine biologist who'd been looking at Ps Ano in its original aquatic environment. Next to her sat George, then Linsey.

"Hi everyone," Ayden said. "I'd like to table a first draft of our preliminary report. You've all contributed to parts of it, but now I'd like you to read through the whole thing, including the areas outside your own specialties."

Ayden nodded to the Professor. "Professor Sherman has found a tetracycline that may work as an antibacterial agent, at least in the short term. We're about to start hive testing.

"My mother's team has produced a modified bacterium that doesn't generate the waxy deposits, but so far, the engineered bacteria is a very poor competitor for Novo. They're trying forced evolution, breeding a robust version by stressing the population until 90 percent is dead, then allowing the survivors to proliferate before starting another stress cycle. It may take some time though."

Cate nodded at Ayden, still beaming.

"Robyn Cassidy's team has found a louse which appears to feed on the waste of the original aquatic bacteria, and in time, it may be possible to adapt the louse to a non-aquatic environment.

"As you all know, I've been working on an idea, too. Most of you have helped me with bits and pieces of research and testing but I haven't really explained in detail what I've been working on. It looks like the original fairy shrimp bacteria, Ano, is preyed on in its natural aquatic environment by a bacteriophage, a virus called T9, which has a non-aquatic relative called T12. Initial testing indicated that the modified bee bacteria, Novo, was immune to T12 because of the shape of T12's protein coat. Fortunately, the protein is determined by a single gene so we've spliced in T9's. In the lab, Novo succumbs to T12x as we're calling it. We need to do a lot more testing, of course, but we have reason to be optimistic. T9 is disastrous for desiccation resistance of fairy shrimp eggs."

There was a lengthy silence, then George spoke. "You've identified a candidate bacteriophage virus, tested it on the target bacteria, isolated the cause of the failure of that test, determined the solution, and successfully spliced the requisite gene into the virus—in one month?"

"We've got a pretty good team here, Dr. Reyes."

"Right," George said, shaking his head with a smile. "How long till we have the results of tests in situ? On the bees?"

"Professor Sherman?" Ayden said.

"I'd say we could give you preliminary results in another month, following which we'll need to conduct field trials."

The virtual fog got even heavier as Professor Sherman shifted in his seat. Ayden couldn't resist. He flicked into console mode, found the meeting comms links, and snuck a peek at Sherman's original signal. The professor was sitting at his desk looking at his desktop screen, which, no doubt, was split into separate windows so he'd miss the whole VR experience. Ayden smiled and flicked back to the virtual meeting.

"Okay," George Reyes said. "Comments on the draft by end of tomorrow, please. Next, we need to start working up a detailed risk analysis for the field trial proposal."

The meeting continued, focusing on the minute detail of experiment design and risk management, but Ayden frowned and flicked back into console mode. It listed six comms links, one for each of the attendees but when Ayden had looked at the list before, he was sure he'd seen seven.

Someone had hacked into their meeting.

Wednesday, June 29th

Greg Fanshaw was at the edge of the orchard, leaning over something mechanical resting on the open tailgate of his truck. His dark blue shirt was coated with dust. A broad-brimmed hat shaded his dusty, lined face.

"Good morning," Ayden said, as he got closer.

Greg turned and, seeing Ayden, put down the wrench he was holding and started to offer Ayden his hand, then realizing it was covered in grease and dust, withdrew it with an apologetic smile.

Ayden still offered his hand, and after a moment's hesitation, Greg shook it, his smile broadening.

"Good to see you, Ayden."

"Irrigation pump?" Ayden said.

"Replacing the seals. Can't afford to lose a drop."

"Mrs. Fanshaw says to come in for morning tea. She's baking something that smells unbelievable."

"Helen does the best blueberry muffins you'll ever taste. The pump can wait."

As they walked between rows of trees, Ayden eyed them. They bore some fruit but not much. Greg must have followed his gaze.

"Not even enough to pay the bank's interest bills," he said. They walked in silence for a few moments. "So what brings you down here?"

"I was in the area. Thought I'd stop by and see how you were getting along."

"In the area?" Greg raised a bushy eyebrow.

"I'm visiting Sino-Pac Horticulture this afternoon. They have an experimental late-blooming blueberry crop in a twenty-thousand-square-foot greenhouse. It's the perfect setup for our stage-one field trials."

They'd nearly reached the farmhouse. Greg Fanshaw stopped. "Field trials? You got us a fix?"

Ayden stopped and turned to him. "We've modified a virus to kill the bacteria. In the lab, the bees are back to collecting pollen within days."

"Maybe there's hope after all."

"If I can get Sino-Pac to agree, then I think I can persuade the EPA to let us start enclosed field trials early. We need to test our virus in contact with a complex natural ecosystem of microorganisms."

"But it'll work? Your virus will kill the bacteria?"

"It'll definitely kill the bacteria. What we need to do is make sure it won't interact in some unexpected way with other viruses or bacteria, swap DNA with them, and cause a bigger problem than we're solving."

"You can do it in time, though? For next spring?"

"There's months of testing and analysis to do, but if we get started fast enough, and if all goes to plan, then yes, you can look forward to a commercial crop next year."

"It's becoming clear to me, Ayden, that if anyone can do it, you can. Who are you meeting at Sino-Pac?"

"The local manager, Jeff Tan. Regional management says it's up to him."

"I know Jeff. He's a good man. I'll call him."

"Thanks, that'd be helpful."

"No. Thank you. I'm told you were behind the financial assistance package, too. I wouldn't have applied for it myself—no Fanshaw has ever asked for charity 'til now—but I'm glad to have it. It's all that's keeping the bank at bay."

They started walking again.

"It's not charity. Can you imagine what it'd do to Oregon's economy if it allowed its primary producers to go broke? Believe me, it's cheaper for them to help you keep your head above water for a while. It's an investment."

"Hmmm…I have to admit, when we met, I underestimated you considerably. I want you to know, I'm grateful for all you're doing." He turned toward the house and inhaled deeply. "Yep, blueberry muffins. Give me a minute to get cleaned up, and I'll see you on the porch."

Ayden washed his hands and was heading toward the porch when a photograph hanging in the corridor caught his eye. It took him a moment to recognize the couple holding their baby. They looked so young and unblemished by cares.

He heard Greg speaking quietly in the kitchen. "Looks like I might not be needing you after all, old girl."

Then the sound of Greg's footsteps told Ayden that he was heading out the far door of the kitchen.

As Ayden passed the kitchen door he glanced in. It was empty. Greg had been in the kitchen by himself. For some reason, Ayden's eyes went to the double-barrel shotgun hanging on a rack high on the wall.

W.R.R. MUNRO

8 :
DECEMBER 2033

There was no commentary accompanying the clip, and no audible conversation among the hundreds of people who stood and shuffled. No one looked at the camera, and they didn't seem to be making eye contact with each other either. A light breeze could be heard occasionally playing with a dry leaf or a scrap of paper or a discarded plastic bag. Whoever shot the clip was walking alongside the stationary queue which wound its way along the street, around the corner and finally into a bakery.

A blackboard sign stood on the pavement, leaning against a lamp-post. Scrawled in chalk were the words "Pain du Jour—Prix Contrôlée" [Bread of the Day—Controlled Prices]

THE SOCIAL SPECTATOR, "EUROPEAN FOOD PRICES ROCKET AGAIN. WHAT IS TO BE DONE?" ENG.SOCSPEC.VITRE.FR

Thursday, December 1st

[I'm told you're doing magic, Ayden. Well done.]

[Thanks, William. I think we've got everything we need to get it over the line.]

[I'm sure of it.]

Wednesday, December 14th

The EPA publically convened to assess the field trials and determine whether the intervention should be approved and funded.

The hall could easily have accommodated a hundred people or more, but most of the tiered seating was empty. There were, perhaps, a dozen people physically in attendance, easily outnumbered by the ten board members, their assistants, and George Reyes's team. They were, however, registering over a hundred thousand virtual spectators.

Ayden's presentation was succinct but accompanied by gigabytes of supporting data. There was a single submission against the proposed intervention, ironically from one of the more radical members of BioWatch, who wasted no time in denouncing Ayden. Ayden thought he raised some valid concerns, but the EPA Board simply ignored him. They voted immediately and unanimously to approve the intervention as proposed.

As the hearing closed, Ayden's phone informed him that Franklin Hawthorn, the chairman of the EPA, wanted a quick word with him. Fittingly, the executive offices were a short walk from the public hall, through a courtyard of bare-trunked ornamental maples framed by walls of green conifers.

At Hawthorn's outer office, Ayden was greeted by a secretary who ushered him in.

"Thank you for dropping by on such short notice, Ayden, and of course for your exemplary work on the project. Please take a seat. Now, can I get you anything? Coffee, water, scotch? Champagne, perhaps?"

They sat in aging but comfortable chairs in a corner of Hawthorn's large, burgundy office.

"Water, please, and thank you for approving the intervention, Mr. Hawthorn."

"It's what we do, Ayden." Hawthorn turned to his secretary. "Mary, the usual for me, please, and a water for Mr. Walker here." Hawthorn looked to be in his late sixties or seventies. Mary was probably in her late forties, trim and attractive in a restrained way. She fixed a drink for Hawthorn. It might have been scotch and soda, equally it may have been ginger ale. Handing Ayden his water and Hawthorn his drink, she sat down at Hawthorn's side and watched Ayden.

"You're wondering how an old coot like me can keep up with all the science that gets thrown at him." It was a statement rather than a question. "Well, I'll let you in on a secret. I don't. But I do have a finely attuned 'bs' detector, and I know someone in almost every field I can rely on for an honest, expert opinion. Now, I'm sure you want to go out and celebrate so I won't keep you, but I wanted to meet you informally and offer you another project. We're trying to clean up a mess in South Carolina. As of about four months ago, Hunt River started producing all sorts of awful readings for PCDDs." He paused, casting a glance at Mary.

"Polychlorinated dibenzo-*p*-dioxins," she said.

"Thank you, my dear," he said warmly to Mary before focusing again on Ayden. "We've conducted an investigation, and it seems an old pesticide plant about five miles from the river is the source. They were shut down in twenty-two, and their waste-holding tank was cleaned up, but it now looks like the damned thing had leaked for years and the extent of the seepage had never been properly investigated. An underground PCDD plume has been seeping toward the river ever since.

"This is particularly disheartening for us. Hunt River was one of our success stories. Back in seventeen it was in terrible shape, but we brought it back to excellent health—until now."

"How can I help?"

"We've had a barrier dug in, which should hold it for a while, but we need a permanent solution. Right now, we're looking at a proposal to bioremediate the plume using recombinant micro-organisms. The proposal appears solid, but the risk evaluation looks a bit too smooth for my liking, and there are questions about the independence of the authors. I want you to dig through the proposal, let me know what you think."

"I'd be happy to, Mr. Hawthorn."

"Call me Franklin, and reserve any statements about happiness until you see their fourteen gigabytes of supporting data. They've certainly assembled a weight of evidence. Makes me wonder what they're trying to hide under the pile."

"I don't mind digging."

"That's what I wanted to hear. Now, here's the rub, these guys have some influence, and Hunt River is a political hot potato. If we knock back their proposal, we're going to need a viable alternative."

"Ahh. Well, I can't really promise anything, but I'm happy to have a look."

"That's all I can ask. You come very highly recommended, and your bee intervention proposal was exemplary."

"I'll do my best."

"Good." Hawthorn turned to Mary, who had been silently examining Ayden since she sat down. "Mary, please have legal draw up a contract and get in touch with Ayden here."

"Yes, sir."

"And, Ayden…" The old man locked eyes with Ayden.

"Yes?"

"I'm assured I can rely on your discretion regarding some of the matters we've discussed this evening."

"You can."

"Excellent, now go out and celebrate."

As Mary showed Ayden out, her phone pinged Ayden's with contact data for Hawthorn and herself.

"I'll email you a draft contract and initial budget in the morning. Naturally, the contract won't mention anything about alternate proposals, but if at any time you need resources, additional budget, whatever, just let me know."

"Thank you." Ayden left, certain that Mary was one of the people Hawthorn relied on. On the way to the restaurant to meet the others, Ayden ran a quick search on the Hunt River project. There were a few blogs suggesting that Professor McKinnon, who was leading the project, had numerous business dealings with SL Remediation, the supplier of the genetically modified bacteria they were proposing to use.

• • •

George Reyes invited Marc, Cate, and Ayden to dinner in a French bistro with candles and menus printed on paper. Ayden's menu didn't list prices, and when he tried to order French onion soup, George insisted he have a steak.

"We have much to celebrate. Besides," he said with a wink. "I have an expense account."

Ayden described his meeting with Hawthorn and mentioned the possible connection with SL Remediation.

"Well, my boy," George said as Cate beamed with pride and Marc shook his head with a rueful smile, "you never cease to amaze me. First you drag Sherman into the field so he uncovers the problem, then you find the solution. Meanwhile, you make first an enemy of, then an ally of Genenco, and now you're being courted by the EPA, being invited to join the inner circle by the look of it."

"Inner circle?"

"A small group of close advisors who help the EPA evaluate proposals. Men and women of significant influence." George's tone grew serious. "Power. Money. Big fish with sharp teeth. Be very careful of what you say, and who you say it to. In other words, watch your back."

"Or just say no," Cate said. "Walk away from the role. You don't need it. You don't need enemies, and you really don't need the spotlight. Stick with science, it's cleaner than politics." Cate smiled apologetically. "Sorry, George."

"Not at all. You're quite right," he said.

"Don't worry, Mom," Ayden said. "I've got zero interest in politics, but I am going to accept the role. It sounds fascinating, and it's got to be done right. It's not only Hunt River's ecosystem at risk, it's all the downstream ecosystems, too, but my input will be strictly scientific. I won't be influenced one way or the other."

"Hey, pity the politicians," Marc said smiling. "They don't stand a chance."

"You know, I believe you're right," George said. "Ayden's not the only one who'll need to take care." He turned to Cate. "And if thoughtful, incorruptible people refuse to touch anything political, where will that leave us?"

Friday, December 16th

Ayden peered into his specs.

Jaal's avatar was vaguely lizard-like in form, recumbent on impossibly deep, purple moss, long tail occasionally shifting lazily, but the lustrous blue-black skin and powerful musculature looked somewhat human.

The environment was equally rich in detail. To either side, near-vertical walls of rock and fern climbed into the sky and ran with water. On one side, a massive u-shape was cut out of the mountainside a thousand feet overhead. Every surface was coated with a million diamonds as droplets of water caught the light of two bright suns.

"Glacial?" Ayden said, looking up at the hanging valley.

"How can I be of assistance, Ayden Walker?" the avatar said. Its mouth-parts were complex and entirely alien but moved realistically with its speech. The avatar's eyes, however, were black and impenetrable. Jaal obviously had no interest in small talk. It had politely refused to supply any information about itself. Ayden had no idea of Jaal's age, gender, or geographic location. It had, however, come highly recommended.

Ayden had never employed a data miner before. He'd mastered advanced search syntax as a child, and as the Web evolved from a repository of documents into an application-serving, distributed processing grid where most information was contained inside virtual realities and collaboration spaces called virtualities, Ayden's skills had grown. Now, however, the emphasis was on speed and Ayden needed to leverage his capacity.

At precisely nine in the morning after Ayden's meeting with Hawthorn, his assistant, Mary, had sent through a contract specifying a day-rate more than double anything Ayden had earned before. He had signed the contract electronically, letting his specs perform a retinal scan while providing a voice sample and entering his PIN. Twelve seconds later, Ayden had unrestricted access to the complete Hunt River proposal. He would comb through the proposal himself, but he couldn't trust it to be his only source of data. He needed to survey the field for alternate approaches.

The contract included a healthy budget for out-of-pocket expenses, and Mary's message had reinforced the need for speed so Ayden had posted a query to SciVirt asking for recommendations.

A day later, Ayden had one freelance miner on board, Sheri Tallis, an ex-journalist and single mother in Adelaide, Australia. She had promised to provide a comprehensive report into the relationship between Professor McKinnon and SL Remediation. Now he was briefing a second—Jaal.

"You read the summary I sent through?"

The lizard head nodded.

"On the face of it, it doesn't look controversial. Genetically modified bacteria have been used to remediate toxic spills and waste for decades. Amazingly enough—I only discovered this yesterday—they started using microbes to clean up oil residues halfway through the twentieth century."

"George Robinson in the 1960s, yes, and the recombinant DNA technique was engineered in 1973 by Stanley Norman Cohen and Herbert Boyer to create the first transgenic bacterium," Jaal said.

"Ahh, yep, okay." Ayden smiled ruefully, not sure whether Jaal was being competitive or telling him to get on with it. "Well in this case, the PCDDs are dissolved in a solution that doesn't mix well with water so I'm worried the modified bacteria, which are waterborne, will be limited to the surface of the toxic plume. The proponents make much of the fact that their bacteria has been successfully used before with dissolved PCDDs, but when you link through to the actual projects, you see that previous successes were all 'pump and treat.' Contaminated soil and water were dug and pumped out of the ground and treated in vats that could be agitated, mixing the bacteria with the toxic solution. In this case though, the contamination is too spread out so they're planning to inject the bacteria into the ground."

"How can I be of assistance?"

Ayden stared at the avatar as it suddenly occurred to him that he might be talking to a sophisticated artificial. Artificial constructs were generally used by corporations seeking to minimize costs, but they'd also been adopted by individuals with Asperger's or physical disabilities or simply because they didn't like being interrupted.

"I need to be confident I haven't left any stones unturned. Appendix three claims to list and consider all of the alternative remediation options available. I'm looking for independent confirmation that there are no alternatives they haven't mentioned."

"I don't envisage any difficulties."

"You'll need to dig through university databases, science virts, and com spaces as well as doing a general document search."

"Of course."

"And if you do find anything they haven't covered, I'll need a comprehensive report, and I'm sorry to push, but I'll need at least your initial findings as quickly as possible."

"I have a job I will need to complete before I begin."

Ayden winced. He'd hoped to at least know if there were other alternatives within the week.

"You will have my initial findings in forty to fifty hours. The timing of a comprehensive report regarding the alternatives, will depend on how many there are."

If he could deliver as promised, Ayden didn't care how unusually Jaal chose to present himself.

He was becoming very, very busy.

Tuesday, December 20th

He watched the bacterium wriggle. Its body was oval-shaped, but fatter at one end, like an egg only more so. Its flagella, each longer than its body, whipped viciously. He moved closer. Fine hairs on its body caught misshapen lumps, and they were absorbed within the cell. Closer. Ribbons inside the bacterium wrapped around the PCDD and literally tore it apart. The bacterium ate toxic PCDD and expelled harmless metabolites. Commonplace perhaps, but a wonder to watch and contemplate.

[Text message from William Hanford,] his earpiece said.

Ayden continued to watch, visually absorbed within his small but growing HUNT virtuality. [Read it,] he subvocalized.

[Xenith is worth a look.]

Ayden had no idea who or what Xenith was. He'd look them up later.

After four days and well over a hundred hours of intensive analysis, Ayden was convinced the proposal would protect the Hunt River from PCDDs. There remained, however, a question surrounding the metabolites. Though they were considered generally harmless, Ayden had turned up concerns raised by biologists from the National University of La Plata in Argentina that amphibian reproductive cycles might be affected by some of these metabolites. Their paper hadn't been peer reviewed yet, but they'd been happy to provide Ayden with a draft.

He began to read.

[Incoming voice call from Professor Sherman,] his earpiece said.

[Accept.] "Hi, Professor."

"Ayden, my boy, I'm not disturbing you I trust?"

"Not at all. Thank you for having a look at my monitoring plan."

"My pleasure. It's good work. I have no comments other than to note the amount of workload you've allocated to yourself. As I recall, the EPA is only paying you for a day a week to monitor the bee intervention, whereas I estimate you've given yourself at least twenty hours per week of work."

"Thanks for your concern, Professor, but I'll cope."

"I've no doubt you would, and frankly I wish more young people today adopted that attitude, but I understand you're under tremendous pressure with Hunt River, not to mention a suddenly frantic social schedule?"

Ayden laughed outright. "Yep. Suddenly the entire science community wants to meet me."

"Of course, they do. The EPA's latest weapon is a dark horse, and the local establishment is in a flap, desperate to have a look-see, place a few hooks. My spies tell me you're being wined and dined all over town."

"I'd no idea how many expensive restaurants Seattle has. Thankfully, Linsey's gate-keeping for me, keeping it to a maximum of one a day and pushing most of them to breakfasts and lunches."

"Good tactic. Keeps them short. Hawthorn'll be loving it. The old codger never misses an opportunity to stir things up. But listen dear boy, I think I can take some of this off your plate. I'm a firm believer in giving undergrads something real to chew on, and it strikes me that, heavily supervised and doubled up, one of my classes could take on some of the more mundane aspects of this job."

Ayden flicked HUNT away and dived into BEAN. He and the Professor quickly agreed on the key details for the latest U Dub School of Entomology class project, then he was back into HUNT and the La Plata paper. He was able to work for nearly an hour before the next interruption.

[Voice message from George Reyes.]

[Play message.]

"Hi, Ayden. I've had a look at your draft contingency plan. I have no problem with its substance, but bearing in mind it'll be a public document, I think we need to finesse the language a bit. So as not to play into the hands of alarmists. Can we schedule half an hour sometime over the next couple of days?"

The point of contingency planning was to envisage and plan for the worst possible disasters, and when releasing genetically modified micro-organisms into the wild, the possibilities should scare people. Ayden sighed. By "alarmists," George meant groups like BioWatch, and Ayden knew better than most how some of the gang would sensationalize his document.

[Voice message to Linsey Carr,] he said.

Ayden asked Linsey to schedule a conference call almost unconsciously. His mind was already back on the La Plata paper. Their science looked robust, and they cited a few other papers which also linked the "harmless" metabolites with amphibian reproductive problems. Ayden's hands moved. Flick, out of HUNT. Point, into SciVert. He composed a post, looking for an amphibian expert to give him a second opinion on the paper. Flick, back out of SciVert, into the advanced interface for FUZZY, Ayden's search bot of choice, looking for surviving amphibian populations in the Hunt River. In no time, he was completely absorbed in HUNT again.

Eventually, he surfaced. His eyes were tired, his neck stiff. It was after ten p.m., enough for the day. He glanced at his inbox, at all the messages and calls that hadn't made it through his filter. Nothing urgent. He scanned his daily memo from Linsey. Mainly scheduling items, a time to meet with George Reyes about the contingency plan, two more lunches next week, and a proposal from Xenith Foundation. He was invited to join the grants committee. They provided grants for research into the human brain and particularly the phenomenon of consciousness. It sounded interesting, but Ayden wasn't sure he had enough hours in the day. And he wondered if William was behind it. There was a link to an audio summary, so Ayden decided to listen as he headed home. At a brisk walk, he could get there in forty-five minutes, feeling a lot fresher.

He emerged from the building and paused to watch the torrent of humanity bustle past, ignoring the beggars who sat with their backs resting against the base of the office tower. Among them sat a weathered but young woman with a small child in her lap.

A gust of wind brought him a waft of cooking smells from the roadside kiosks and Ayden realized how hungry he was.

[A rainsquall approaches. ETA 3 to 5 minutes,] his earpiece said. Ayden didn't have a coat or umbrella with him. So much for walking. He summoned an auto-cab and walked over to a kiosk to buy a cup of ramen. He bought two extra cups and gave them to the young woman. "Rain coming," he said. She thanked him and scurried off, child in tow.

He was halfway through the noodles, sheltering from the rain under an eave with perhaps fifty other people, when the little two-wheeled electric rickshaw arrived.

[Home,] he instructed and ignored its progress as it threaded its way through the variety of vehicles that had emerged when cheap, reliable collision avoidance technology rendered expensive, heavy "crashworthiness" safety features obsolete. The little cab's gyroscopic balance even meant he didn't need a lid for his noodles.

Ayden was nearly back at the condo when Tess called.

"Who are you sleeping with?" she said in mock exasperation after Ayden had described the Xenith Foundation proposal. "And does he or she like girls, too? Next you'll be a home page item on worldscientist.sci. In fact, I've plotted your rise to prominence, extrapolating from recent trend data, and the curve is parabolic. You'll win a Nobel Prize next year, become president of the United States a month after that, and global dictator the following week. It'll take you two years to dominate the Milky Way galaxy but only six months for the rest of the universe to fall into line—immediately following which it will belatedly be recognized that you are, in fact, God."

"Or you could do my PR and we'd get there in half the time."

"Seriously, Ayden, you've gone from obscure field researcher to science administration's golden boy. There's something you're not telling me."

All the levity left Ayden's face.

"Actually, there is a lot I'm not telling you at the moment, Tess."

"Wow, now I really want to know what you're talking about. What'll it take, Ayden, 'special massage with happy ending?'" Tess said, jesting half-heartedly, as if she were suddenly wondering whether she really wanted to know.

"In any case, I'm sure the one impostor will soon enough be replaced by the other."

"The what?"

"Rudyard Kipling. 'Meet with triumph and disaster and treat those two impostors just the same.'"

"Oh, he even quotes poetry, very impressive. That's the 'you'll be a man' poem, isn't it?"

"I doubt anyone has quite met his standard, but I suppose we could all do with a target."

"So you think your shooting star will fall to earth?"

"Murphy's Law is bound to kick in at some point."

Wednesday, December 21st

The next morning, as Ayden was about to enter the project office building, he paused for a moment, closed his eyes, and turned his face to feel the warmth of the sun through the cold air. There was no space at the base of the building for a garden so it met its landscaping obligations with hanging gardens of greenery, which draped over its steel and glass façade. Lightly dusted with morning frost, it was quite beautiful.

When Ayden opened his eyes, he found a middle-aged man standing in front of him. He wore a fashionable, v-neck, knitted shirt and matching jacket, and his hair was the unnaturally intense dark brown that comes from a bottle, but his pallid skin and the bags under his eyes gave him an unhealthy air.

"Walker? Ayden Walker?"

"Yes?"

"I'm Winston Lowe. My name mean anything to you?"

"You patented Rhino Mozzies. I tried to find you. I have questions for you, but you've sought me out. What can I do for you?"

"Not here. In my car."

He gestured toward a two-seater REVA Dart. Its doors un-locked with a click.

"Drive mode. Manual," Lowe said, doing up his seat belt. As soon as Ayden did likewise, the car lurched forward silently. Evi-dently, Lowe had already put his foot on the accelerator so that the car was only waiting for the seatbelt to be engaged before proceeding. Within four blocks, the little car's collision avoidance system had been forced to apply the brakes twice. Ayden didn't know whether Lowe always drove this badly or whether it was a result of his obvious stress. Perhaps he was even deliberately driv-ing dangerously to unsettle Ayden.

Ayden was glad he didn't live in an era when cars relied on their human drivers to avoid hitting things.

Lowe didn't look at Ayden, even when the car was motionless in the traffic. "All right, who do you work for dammit?" Lowe said.

"I've been appointed by the EPA to—"

"Cut the crap. Who do you really work for?"

"That's really who I work for, Dr. Lowe. My job is to solve the bee pollination problem. Actually our solution has already been—"

"Approved. I know. What I want to know, is why you aren't trying to find out how the bacteria transferred from *Toxorhynchites Rutilus Septentrionalis* to *Culex Pipiens*."

"Technically, it's outside our brief, but I'd certainly like to know whatever you can tell me."

Lowe hauled the car over to the side of the road. "Shut it all down and strip it off," he said as he shut down his own phone gear and dropped glasses, cuffs, neckband, and earpieces into a compartment in the center console between them. He held the compartment open until Ayden complied, then pulled the car out from the curb with a lurch of acceleration.

"You want to know what I can tell you? I can tell you they're a bunch of thieving whores. I can also tell you that there's no way my bacteria could have transferred to common bloodsucking mosquitoes by accident. I'm not that sloppy." Lowe paused. "Didn't used to be anyway. I used to do good work. Til they destroyed my life."

He drove in silence for a while. Finally his driving became a little less erratic.

"Dr. Lowe, please tell me who 'they' are."

"They conned me out of my patent. Oh, the up-front was okay. It was enough for me to buy a nice house. I even bought myself a hybrid gas-turbine/electric Porsche. The '25 model—you know it? No?"

Lowe smiled for the first time. "Man that was a fast beast. The electric motor got you off the mark so fast it almost hurt, then the gas turbine cut in and just kept pushing. They had to fit the damn thing with a proximity sensor in the headrests to make sure it was properly adjusted relative to your head…didn't want to face whiplash claims, the thing was so brutal." Lowe's smile dropped away. "Sold it after I lost my license."

Ayden let Lowe ramble on with minimum interruption. Eventually, he stopped the car on the side of the road in the middle of suburbia, and they just sat there talking.

"Thought I'd hit the jackpot, you know?" Suddenly, he was searching Ayden's face as if for understanding.

"The royalties were going to make me filthy rich. Governments would buy massive amounts of Rhino Mozzies 'cause it would be cheaper than dealing with mosquito-borne diseases, and the Joneses would pay top dollar so they could have their garden parties without their guests being pestered.

"And the jewel in the crown? Global warming. Warmer temperatures are kind to mosquitoes so the market would keep growing.

"And I wouldn't have to worry about all the mundane details of regulatory approvals or multinational commercialization. MTC would do all the dirty work and pay me a good percentage of revenues. I didn't know all that much about MTC, but they threw money about like it grew on trees, and apparently they were part of Genenco.

"But instead, they totally screwed me. They gave me a lab and encouraged me to 'continue my research,' but they never launched it. That's what really gets me. It doesn't make sense! Why spend a chunk of money on a patent then never exploit it? There was something screwy going on, but I've never been able to work out what.

"Eventually, I got fed up. Started making noises. They threw me out of the lab. Cut off all communication except through their damned lawyers. When I went looking for lawyers, all the best intellectual property specialists claimed 'conflict of interest.' They claimed to be on retainers with 'business interests associated with MTC Trust.'

"Finally, I found a lawyer willing to take the case, and then it really hit the fan. I got blackballed. I was talking to a couple of universities when suddenly they wouldn't return my calls, so I tried to do the rounds of the biotech start-ups, but MTC had gotten to them as well. Then what do you think happened?"

Ayden had assumed the question was rhetorical, but Lowe was looking at him, waiting for an answer. "Uh, I couldn't begin to guess."

"The goddam IRS swooped, that's what. Out of the blue, in-depth tax audit. Cooked up some reason I owed them money. It was all crap. I ended up winning the case, but the award for legal costs didn't come close to covering my bills."

"Sounds rough."

"Ya, rock bottom, eh? No. Then I met Crystal." Lowe shook his head. "Should have known better. I assumed it was the car. I was still hanging on to the Porsche at that stage. Man, she was sex on a stick. Seriously liked to party. Even as a student, I'd never really done much in the way of drugs, but I had to snort blow and drop little blue devils just to keep up.

"Then one night, really late, she said she felt sick, asked for a lift home. I'd never been there. We always used to crash at my place. Great, I thought. Finally, I get to see her digs. Only she didn't invite me up.

"Then on the way home, I got pulled over by the cops. Claimed my behavior gave them probable cause. Searched the car. And surprise-surprise, the stash of coke they found under the passenger seat was just above the prescribed level, just enough to get me as a dealer. I never saw Crystal again. She disappeared from the face of the earth, and no one by her name had ever rented an apartment in the building where I'd dropped her.

"I finally gave up when my own lawyer told me he had been asked to deliver a message. He had been told to tell me that life could get even worse or it could get better. I'd already lost my job and my car, was looking to lose my house as well, just to pay the legal bills. He advised me to give it up. My own damn lawyer was in cahoots with those bastards. I had nowhere to go. I caved."

Lowe was offered a quiet teaching position and a favorable settlement for his legal bills. He kept his house and had work, and in Lowe's mind, it was clear that the continuation of this new existence relied on him not pursuing MTC Trust.

"So the bad guys won, the world stinks, and I occasionally get lucky when the divorcee next door takes pity on me. Move your knee."

Lowe reached past Ayden to fish around in the glove box. "What you want is in here." He held up a memory chip. "It contains the sequences for my modified *Ps Anostraca* and a detailed analysis of the risk of transfer to *Culex Pipiens*. Compare what's here with your samples. No way could it have happened naturally. I don't know why, but they did it deliberately."

Ayden reached out to take it, but Lowe pulled his hand away.

"Hang on. There's nothing I want more than for someone to expose what they did and work out why, but you seem like a nice enough kid so before I give this to you, I gotta warn you. You give these guys grief, and they won't hesitate to ruin you. So if you decide to do something with it, you need to make sure you've got some big guns lined up behind you. And don't trust anyone. You still want it?"

"Yes."

Lowe handed him the chip and opened the center console so Ayden could retrieve his phone gear. "Don't trust that stuff either. If it's not bugged already, it soon will be."

Ayden's door clicked open.

"Good luck, kid. You'll need it."

. . .

"This whole thing worries me," Marc said when Ayden finished describing his conversation with Lowe. They were in the condo fixing a basic stir fry for dinner.

"Maybe Lowe's just looking to blame someone else for his problems. He claims they broke him, but he can afford registration for a private car."

"But then why didn't MTC commercialize Rhino Mozzies?" Cate said. "And think about what Gardner nearly did to us, how that would have looked to outsiders."

"According to William, MTC accidentally allowed the bacteria to transfer, then pulled the plug on Rhino Mozzies to cover it up."

"So it's Lowe's word against Hanford's."

"Essentially, yes. Which is good news."

"Why?" Cate said.

"Because, once you've had a chance to analyze Lowe's data, you may be able to tell us whether the transfer could have happened accidentally or not, and that will tell us who's lying."

"I can't guarantee a cut-and-dried answer."

"No, which is why…Dad, can I ask you a favor, too?"

"Sure."

"Have you got much going on at the moment, deadline-wise?"

"Nothing onerous. I'm analyzing some genographic data for Eli Lilly, but it's pretty basic stuff and I'm way ahead of their schedule. What can I do to help?"

"We've got genomes from infected mozzies from all over the Americas and parts of Europe, thanks to some BioWatch colleagues over there. This isn't going to stretch your abilities either, but if the mozzie infection was accidental, then it would have originated from a point source in the U.S. then, at some later stage, hopped across to Europe and spread from one or a few sources there as well. Could you look at the mutation rates for me? See if they tell us anything?"

"Sure, but what am I looking for…a little more specifically?"

"If William's version of events is accurate, then the mosquitoes managed to spread very widely in less than four years and different strains of the bacteria would have emerged. I'm hoping they'll reveal the pattern of spread."

"Easy."

"Thanks. One more thing…can we keep this just between us for now? I don't want to raise William's ire for nothing."

"About that, aren't you going against your agreement not to investigate the source?"

"The deal's off if he lied to me about accidental transfer."

Thursday, December 22nd

"Can't prove your Professor McKinnon's doing anything illegal, but I've dug up enough to raise some serious questions."

Sheri Tallis's video window was one of about a dozen he had scattered around his HUNT-scape, but she had his full attention. Her short, dark hair was tussled, and her tanned, lightly weathered face was free of makeup and glowed with health. Her Australian accent reminded Ayden fondly of living there.

"He's been involved with nine remediation projects over the last five years, and six of them used SL Remediation."

"Okay, but that may just mean he's found a good service provider, and he's sticking with what works."

"Except that in three of those six cases there have been problems."

"Hmmm, remediation's a difficult business. Problems are part of the game." He smiled. "You've got something else."

She grinned back. "Sure do. McKinnon has a half-brother, different surname, Cordova. Now, Cordova is a property developer, and he's done several developments with some outfit called IC Investments, and guess what? IC Investments is owned by the same guy who owns SL Remediation. Like I said, it doesn't prove anything, could be coincidence, could be that's how McKinnon heard of SL in the first place."

"But it would also give them a channel for money if they wanted it."

"Ooh yeah." There was a crash in the background. "Damn, that's my youngest in the pantry. Gotta run, Ayden. Here's my report." [link] "And my bill." Quick smile. "Please give me a call if there's anything more I can do." Then she was gone.

Jaal, of course, hadn't phoned. He'd sent a text-only report, and he'd found a possible alternative, phytoremediation.

It never ceased to amaze Ayden how much "modern technology" wasn't so modern when you looked into it. Apparently, the ability of some plants to naturally extract high levels of zinc from the soil was first noted in the 1880s, and by the end of the twentieth century scientists were experimenting with a range of transgenic plants extracting toxic-heavy metals and pesticides.

In the first decades of the twenty-first century, an industry formed around the use of transgenic poplars to clean up toxic spills. They grew fast, were very hardy, and could be managed to ensure the right root depth. They were also happy growing closely planted, so it was possible to create quite a secure barrier. Numerous variations had been developed—trees targeting different metals and organic pollutants.

One particular varietal seemed close to meeting Hunt River's requirements. It was possible that, with a relatively minor genetic tweak, the trees could be made to extract the PCDDs from the soil and deliver them to the leaves for easy collection and disposal.

Jaal had certainly exceeded Ayden's expectations. In the absence of an exact match, he'd found something that could be adapted. That took serious molecular biology knowledge, as well as data mining skills. Ayden couldn't help but wonder who Jaal was, but he wouldn't invade his privacy by trying to find out.

The patent for the transgenic poplar had not, however, been licensed by any of the biotechs. There had been some use of them to clean up PCBs in Africa and South East Asia but no one was further developing the organism.

Ayden needed to find a well-funded biotech lab willing to risk the resources required to try to adapt the trees to PCDDs. With McKinnon's proposal not requiring any development funds, the EPA wouldn't be able to justify putting public money at risk.

He asked William.

[Doesn't fit with anything we're currently throwing money at. What's your time frame?] William said.

[Yesterday. The plume's on the move and SL's ready to send in truckloads of equipment,] Ayden said.

[Hmm, Bond's our best bet.]

[Bond University in Australia?]

[Publically listed. Entrepreneurial track record. And Heinz Imoos, Dean of the Ecology Department has benefited from a few Genenco grants over the years. I don't know him personally, but I'm sure I can arrange a warm intro for you.]

[Magic. Thanks, William.]

[No problem.]

● ● ●

Franklin Hawthorn refused to read Ayden's draft report. "Give me a verbal rundown. If you can't deliver me the key concepts in ten minutes, then either you don't know what you're talking about or there's a problem with your solution," he said in his scratchy voice.

Hawthorn wore specs but used a simple video window rather than VR and insisted Ayden did the same. After Ayden had sketched out his proposal, Hawthorn had only one question.

"All I really need to know, Ayden, is will they be able to modify the trees in time to do the job?"

"Well, nothing is certain, but preliminary analysis—"

"Stop, stop," Hawthorn said, waving a bony hand in the air. "Don't give me claptrap about scientific uncertainty. I've heard everything to be said on the subject far too many times. If we adopt your proposal and Bond doesn't come through, then we've lost six months and we look like fools. The safe thing for the EPA to do is bury your suggestion and go with the proven technology with apologies to the frogs, which have likely got one foot in the grave anyway. You want me to take a risk. Well, you first. Can it be done, and will they do it?"

Ayden thought for a moment before answering. He was sure the trees could be engineered quickly. The only question was, could Dean Imoos's team do it? Well, if necessary, Ayden could get involved.

"They'll get it done."

"Well, if you're right, then you've pulled a rabbit out of a hat for us, and done so in record time. We can afford to wait six months for scientific confirmation because you turned the assessment around so quickly."

"Glad to be of assistance."

"We'll give your draft a once-over, make sure there aren't any surprises. Mary will come back to you with any concerns. Otherwise, well-done, Ayden. Take a break. I imagine you've been burning the midnight oil with both Hunt River and the bee intervention."

But Ayden was nervous. The genetic tweaking required for the poplars should be straightforward, the technology was well-established, and Imoos's researchers appeared competent. Nevertheless, he went straight back to his office to start preparing the project database for Bond University. He didn't want to leave anything to chance.

Friday, December 23rd

Ayden took off his specs and rubbed his eyes. He stood and wandered around the office, stretching his shoulders and neck.

It was dark outside, a cold clear evening, perfect for a refreshing walk home. For the moment, he'd done all he could to get Bond University moving in the right direction.

His earpiece chimed. He suppressed a groan.

[Voice message from Marc Walker,] it said.

[Play message.]

[Hey Ayden. Just letting you know I've made a start. No real news yet 'cause I've hit a bit of a snag with the DNA data, but your mother has promised to go through it with me in the morning so we'll get it sorted. Have it to you tomorrow, with any luck. In the meantime, I've pulled together all the data we have for the presence of infected mosquitoes worldwide. I've plotted infection densities geographically. Not sure how useful it is though.]

Ayden put his specs back on and flicked the link. It didn't look particularly useful. He floated above the world, occasionally reaching out a ghostly hand to spin the globe on its axis. Yellow meant uninfected samples, blue infected, and color intensity denoted data density so grey meant an area where they didn't have samples.

Grey dominated, of course, but there was more blue than yellow. The infected mosquitoes were clearly dominant.

He'd hoped to find one or two areas of outstanding density, indicating sources of the infection but no such luck, there were many points of maximum density.

[*Knock, knock.*] It was the alert tone he'd assigned to calls from Tess. He flicked open a window. "Hi."

"Hi, Ayden. It's that time again."

"That time?"

"When I drag you away from your desk and remind you to pretend you have a life. And don't try to put me off. I'm in the building, coming up in the elevator. What are you up to anyway?"

"Have a look." Ayden offered her a link to join his virtual environment.

"Nice presentation. Your father's work?"

"Yeah. He's good with the bells and whistles. Impresses the clients, he says. Really, he does it 'cause he enjoys it."

"I love the stars. They're actual constellations aren't they?"

"It's worse than that, I'm afraid. The background is completely astronomically correct."

"So if I—" Tess reached out and spun the globe. Ayden could now hear her voice in the room with him as well as via his earpiece. The starscape wheeled. "Cool, though perhaps not vital from a scientific…what…?" Tess suddenly reached out with both hands and drew the planet closer, zooming in on Europe.

"Wait a minute," she said. "That's not right. What have you done with my data? Ayden, what's going on?"

"Your data?" Ayden looked at Tess's avatar then ripped the specs off his face to look at her. "Please tell me exactly what you think you're looking at," he said.

Her specs cleared. "That's my fertility data, the spread of capacitation-related infertility."

Ayden's stomach felt hollow. He stared at her.

"But it's broader than mine," she said. "You've sourced more from somewhere else and…and what do the different colors mean? I don't understand. What are you doing?"

"Tess, that's not displaying a single fertility datum point."

"What?"

"It's showing the spread of infected mosquitoes."

"Oh…oh my God, you can't be serious."

"Deadly."

They stared at each other.

"And we were worried about bees," Tess said. "How could this have happened?"

Saturday, December 24th

[It's Christmas Eve,] Tess subvocalized.

Ayden glanced at the translucent clock that floated top right in his specs view. It was nearly one in the morning.

[So it is. You have plans?]

They were sub-vocalizing so they wouldn't distract Marc and Cate who sat on the couch opposite in the Walker condo's little living room, specs opaqued.

[I was going to have an all-afternoon lunch with some girlfriends, now I'm not so sure. Christmas Day gets busy, huge lunch with Dad's extended family, hordes of cousins descending from all over California, little ones running around, wonderful chaos really, then dinner with Mom and her family—a much more buttoned-down affair." She paused. "It all seems too normal to think about now, though. If we haven't screwed something up, if the mosquitoes are really causing human infertility, it's just, well, I don't know, huge.]

[Sure is, and while I don't think we've screwed up, extraordinary claims do require extraordinary evidence. So while we're waiting for Mom and Dad to finish, why don't we start to put together a list of the microorganisms known to cause a man's immune system to create antibodies that target his own sperm.]

[My turn to impress you. Here.] She offered him a link. [It's a short list though.]

[Good. Shouldn't take long to—]

At that moment, Cate's specs cleared and Marc tiredly dropped his to hang on the cord around his neck.

"It's nothing we didn't know hours ago but I can give you the formal numbers," Marc said. "The European data shows a 74 percent correlation of the presence of infected mosquitoes with human sperm antibodies linked to capacitation-related fertility issues. Mosquito data's really patchy though. I've got an error bar of plus/minus 15 percent."

"What I find really interesting is the genetic drift," Cate said, "or more accurately the lack thereof. Marc's data snag wasn't a snag. It was just telling us something we weren't ready to hear. I've compared our data for Novo to what was on the disc you got from that Lowe character. Obviously, there's the initial change, which allowed Lowe's version to jump to common bloodsucking

mosquitoes, but after that there's remarkably little drift for something that's been in the wild for years."

"What about the initial change? Any indication of how that
happened?" Ayden said.

"I'm afraid I'm going to have to equivocate on that one. Lowe
knocked out a gene for a protein that Ano needs, one that is found
in Toxorhynchites but not in bloodsucking mosquitoes. According to the notes on his disc, he made a couple of other changes,
too, to better adapt the bacteria to a non-aquatic environment. In
Novo, the knocked-out gene is back, in a slightly different place,
so the organism is now adapted to life out of the water but it no
longer needs Toxorhynchites."

"Hang on a minute," Tess said. "Are you saying someone rein-
serted the gene?"

"That's possible, of course, but it's just as likely they accidentally exposed Lowe's engineered bacteria to the original Ano and
the gene transferred horizontally," Cate said.

"Surely that's a bit of a stretch?" Tess said.

"Happens all the time," Ayden said. "Gene transfer is essentially what microorganisms use instead of sex to provide genetic
variation."

"But how does it affect human fertility? Nothing even remotely
like Ano causes human capacitation."

Cate sighed. "I have absolutely no idea."

"It may not even be real," Marc said. "Our analysis is tissue
thin. The apparent correlation could be an artifact of patchy data
or it could be a coincidence."

"Well, let's find out," Cate said "This is just the starting point."
She paused, looking thoughtful. "Having said that, I suggest we
start treating the mosquitoes as a potential source of infection,
and modify our lab protocols accordingly."

"Jack!" Tess said, staring at Cate.

"Jack Cooper?" Cate said. Jack Cooper was one of the post-
grads helping her in the university lab.

Tess nodded. "He used to boast he was cheating the odds, making sure his genes would feature disproportionately in the human race by donating semen once a month. The other guys said he only did it for the free porn and a hot dog. But he was looking really bummed last time I bumped into him, and I asked him quietly what was up. The last time he went in there they rejected him. It was kind of funny at the time. Anyone would have thought his best friend had been killed. He didn't say what the problem was, but I bet it was capacitation. I bet they rejected him because they found sperm antibodies. We can easily find out. I'll ask him to do a test. He'll want to know. I'll call him now."

"At one in the morning," Ayden said.

"Oh. Yeah. Well, I suppose it can wait a few hours."

"Okay," Cate said. "How's this for a plan? First, sleep, certainly for us oldies. Tomorrow, I'll go down to the lab at U Dub, see if I can borrow a little extra time on the sequencers. I think I need to have a closer look at Ps Novo. Tess, as soon as you can get Jack to provide a sample, we'll get an agglutination test started."

She shook her head tiredly.

"I really hope we're jumping at shadows."

Monday, December 26th

Ayden looked around the train carriage. He couldn't see a single pair of eyes. An immaculately presented, middle-aged lady had her eyes shut, as did a young man in a rumpled suit, sleeping or trying to. Everyone else wore specs, even the two kids who couldn't have been more than ten years old, their hands busy in the air in front of them. Even the young couple who leaned against each other and held hands didn't look toward each other or speak audibly, no doubt absorbed in their own very private universe. Heads moved minutely, hands twitched. No one showed any interest in their hundred-mile-per-hour journey.

William had insisted that Ayden meet him in his Genenco office in San Francisco. It seemed a waste of time and money to

Ayden, but William paid for the ticket and the bandwidth on the train was excellent so really he was as productive there as he would have been anywhere else.

• • •

The office was actually a suite of rooms occupying the whole of the forty-seventh floor of the Genenco building. Ayden was escorted through William's sizeable "public" office to an even larger private office behind. Where the public office was all traditional wood panels, oak desk, and plush furnishings, William's private office was massive and intensely minimal. The carpet, chairs, and couches were all an identical charcoal grey so the seating almost seemed to emerge from the floor. The only other color in the room was white. The low table near the couches was a slab of translucent white glass. William's desk was the same. Even the gleaming white walls, free of art or any decoration, looked like glass. Two entire walls were floor-to-ceiling windows, looking out over the bridge and the bay.

The most unusual aspect of William's private office, however, became apparent only when his secretary closed the door behind Ayden. Immediately, Ayden's phone complained.

[Connection lost,] said the soft voice in his ear. [Data access zero. Awaiting restoration of bandwidth.]

Even in the wilderness, Ayden's phone was able to maintain some level of connectivity. There were few places in the world with no grid coverage, probably none in the U.S., certainly none in the middle of any major cities. William must have engineered a Faraday Cage into the room.

"Please shut down your phone." He wore a collarless, white shirt again, this time without the jacket.

Ayden's phone was incommunicado, but even its limited storage capacity was plenty to record a conversation.

"You live like this all the time? This paranoid? Must be fun."

William smiled. "Take a seat." He led Ayden over to the couches. "Genenco operates in an intensely competitive market. Success or failure is determined by intellectual property, and industrial espionage is routine, and that's just the external threats. I spend as much time looking over my shoulder at my supposed colleagues."

Ayden subvocalized a full shut-down order and waited for William to run his scan.

"Thanks, Ayden. I've looked at your data, and I tentatively agree. It's possible the infected mosquitoes are somehow affecting human fertility. But I wasn't kidding about my colleagues; if there's even a suggestion of blame toward Genenco they'll respond aggressively. We need to tread lightly."

"You're saying this is the first you've heard of the possibility?"

"I am."

"So how'd it happen?"

"I don't know yet that it did happen, just that it's possible. Your data's far from conclusive, but I will find out what happened and how."

"And you'll tell me, right?"

"I don't think I want to commit myself on that one, Ayden, not 'til I have a better idea myself of what's going on. Can I offer you a drink?"

"But you'll help us to stop it? Assuming?"

William leaned forward. "Ultimately, yes, of course. But, actually, I'm not sure that a little delay would be such a bad thing. A brief pause in population growth."

"What? You can't—"

"No, hear me out, please. Let's decide this together. It'll be our first test…can we make difficult, rational decisions together? You think the link's real and that we should stop it ASAP. Very well, for the sake of the discussion, I'll accept the premise that it's real and take the contrary position." William smiled. "Well? Are your debating skills up to it?"

William's enthusiasm was infectious. Ayden relaxed back into the couch. "Take your best shot," he said with a smile.

"First, a lot of men appear immune, over half actually if we take the figures at their face value, so we don't have to worry about the human race disappearing."

"Subject to confirmation and subject to the thing not mutating into something more dangerous."

"Of course. Let's assume that for now. Second, it doesn't cause sterility. With effort, certainly with medical intervention, the infected minority can still reproduce. Maybe it'd be better if everyone was affected. Wouldn't the world be a better place if there were no unwanted babies, if it always took planning and commitment?"

"Planning and commitment? Yes, great. Invasive medical procedures? Not so good. Hang on–" He paused. "That raises an ugly possibility. The infection could be deliberate. Could the motive be commercial gain? Simple greed? No offense, but Genenco has a major investment in IVF clinics."

"None taken; there are some ruthless people in Genenco. But no. IVF uses significant resources perhaps, but it's a relatively small contributor to the bottom line. It's largely a philanthropic investment."

Ayden just looked at him.

"Okay," William said with a smile. "I concede it's at least marginally profitable and excellent PR and a great source of genotype and phenotype information. But, even if volumes rose tenfold, it wouldn't be our largest profit center. I'll look into it, but I doubt commercial gain is behind this. Can we accept that, provisionally?"

"Yes."

"Good. It's a side issue anyway, not central to our decision. Third, most importantly, the elephant in the room. Halfway through the last century, the population of three billion people was starting to stretch the planet's resources. By the end of the century, it was six billion, and it would have taken over one and a quarter earths to support us sustainably. It's now nearly eight and a half billion, and despite an unprecedented international effort to reduce our impact, it would take nearly two earths to support us sustainably. Every lifestyle cutback, every clever solution we implement, is swamped by our growing numbers. We're teetering on the edge of managing climate change, we're forced to adopt risky geo-engineering approaches, we're turning the land surface of the planet into a dangerously limited set of agricultural monocultures, and we're fishing marine vertebrates to extinction, turning the oceans into jellyfish soup."

"Humanity has spent the last two hundred years proving Malthus wrong, but you think he was right after all?"

"Malthus was wrong in almost every detail. He radically underestimated human ingenuity, but at the broadest conceptual level he was right. He was right about human fecundity in the absence of war or famine, and he was right about the planet's finite resources. That we've stretched those resources further than anyone would have thought possible doesn't mean we can go on doing so forever."

"But the population is leveling out naturally."

"The latest U.N. estimates put maximum population at nearly ten billion in 2075. You're an environmental researcher…can the biosphere handle yet another one and a half billion?"

"Not in any state of health. Not if we all keep trying to improve our material living standards. Biodiversity and food webs everywhere are crashing as it is, and the nitrogen cycle's in crisis. No. It's the same as climate change—we're already teetering on the edge."

"It gets much worse. Those forecasts assume gradual increases in longevity. They don't take the biotech revolution into account. The medical advances that we've already seen over the last couple of decades will blow those forecasts out of the water but, of course, that won't show up in the stats until it's happened, until it's too late."

"What sort of numbers are you talking about?"

"The modeling is uncertain because there isn't enough data yet, but there's a very plausible study that estimates eleven billion in 2075, rising to twelve billion by 2100. I'll send you the data but, bear in mind, these numbers are well inside the UN's high variant modeling."

They looked at each other for several moments. Neither man had moved since they sat.

"That'd be disastrous. No question," Ayden said.

"And none of those projections allows for the possibility of our little secret getting out."

"But our cellular telomerase production was engineered into our DNA. That'd never get approval."

"Wouldn't it? Even though they'd see no adverse affects in us, decades on? And even if you're right, scientists have been playing around with drugs that boost telomerase production for decades. I'm watching very closely, and I can tell you that progress is being made. What do you think will happen to human population estimates when they manage to commercialize a drug with the potential to suddenly boost life expectancy by 20 to 50 percent?"

"I shudder to think." Ayden sighed. "Okay, my turn. I concede that human population is a major problem, which has to be addressed urgently. But to engineer infertility into the human race would be a crime against humanity, so to stand back and allow it to happen—whether the vector was natural, accidental, or deliberate—would amount ethically to the same thing."

A look of distaste crossed William's face. "You're not even trying. Fuzzy arguments using emotionally laden clichés won't cut it. You have a point, but you're not arguing it rationally."

"Okay." Ayden thought for a moment. "The urge to procreate is the third-strongest instinct a complex organism has after avoiding immediate life-threatening danger and hunger. According to my parents, the joy of parenthood is unmatched by anything else life has to offer. This infection may steal that from hundreds of millions, probably billions of people if we don't stop it, causing massive human suffering."

"Thank you. I accept that if this infection is real, it will cause suffering, but that's only half the story. Note that you had to qualify your statement. 'After danger and hunger' you said, quite rightly. Unfortunately, death and starvation constitute the other half of the story."

William flicked a long finger.

The walls and windows exploded into a riot of color and motion. Even what Ayden had taken to be windows were, in fact, high-resolution wall screens. William's entire office was an immersive VR room. Everywhere Ayden looked he saw reports on the current state of play in different parts of the world. As Ayden glanced at a scene, he heard the audio for that particular scene. William must have slaved the audio to an eye-tracker following Ayden. Probably the sound was also being directionally projected too—so that only Ayden clearly heard the audio for the report he was viewing.

There was nothing new. Drought and nasty little water wars across Africa. Deadly heat waves in Europe. Famine and food riots across Asia, as crops failed without their annual monsoon for the third year running. Massive unemployment and ever more radical political movements in the U.S., driven by decades of economic malaise. Rampant, unchecked development and environmental catastrophes in India. Nothing new, perhaps, but still shocking to view the images, larger than life, all around him.

"Now you're the one trying to push emotional buttons."

William smiled. "Sorry. Couldn't resist. Here are the facts. This year, millions of people will die horribly after short, unsatisfying lives, and billions live in what you or I would consider appalling conditions. Conditions which arguably cause much more suffering than an inability to have children. Next year, and every year after that, it'll get worse. And it will continue to get worse for decades as our numbers march skywards. Much of the 'developing world' isn't. Progress has stalled, and a great many of children are not planned, not wanted. They are born because of religious objections to contraception or because ignorance, poverty, or sheer male selfishness doesn't allow it. And if somehow we could address the standard of living of the poorest people on the planet, what would happen? Those improved lifestyles would consume even more resources. We'd need four earths instead of two. So what absolutely must happen? What must occur to reduce suffering for billions of people alive today and countless more billions of people in the future? We must reduce absolute population levels in most parts of the world urgently and move to zero population growth everywhere. There are no other options."

"Sure there are. The earth could easily support twelve billion people. We just have to each consume less resources than we do now. It's a choice, a pretty simple one really. Less babies or a lower material standard of living. Point is, the human race has the right to decide. We should use your resources to lobby governments. Educate people. Give them the facts you've just given me, rather than taking away their freedom to decide for themselves."

"That's naïve, romantic nonsense, Ayden. Be honest with yourself. Scientists have known the facts for half a century. Humanity's had every chance to decide, but has proved incapable of doing so. It hasn't helped that economists have stressed the need for increasing population to drive economic growth and the dangers of the aging demographics that accompany lower birth levels. The masses demand the impossible: 'fix it without affecting my lifestyle.' The politicians who tell them the truth—that it can't be done—get voted out. The politicians who promise the impossible get voted in. So you tell me the answer, Ayden. We agree that consumption of the world's resources must be constrained and that it must happen now. So I ask you, what would happen to the politicians who introduce legislation massively taxing consumption or limiting people's right to have babies? Even China gave up on their one-child policy."

"Okay, they'd get dumped. If we could turn the clock back fifty years, I'd say 'educate the children.' Teach them to live more sanely. To examine their own drives more closely."

"But..."

"But even assuming that were plausible, the five year olds we start educating now won't be decision makers for another three to five decades, and we may not have that long to avert disaster."

William was silent.

Ayden spoke again, "I don't like your answer, but I don't have an alternative."

More silence.

"…which is your point, I know."

"What would the most rational response be, Ayden? You and I can expect to be alive in 2100, perhaps in 2150 or even 2200. If we don't do something, what sort of world are we going to be living in? How much suffering will we be witness to?"

William's eyes bored into his. "What would be the gentlest, most compassionate way to reduce population pressures? How could it be achieved without taking a single life?"

"It sounds like you've already decided."

"Not quite. It'd be a big call, but put it this way, I don't dismiss it out of hand."

Ayden was silent for a long time. It was impossible to know if William was telling the truth. Had he already known about the mosquitoes? Were they really an accident?

"What about the risk?" Ayden said. "We have no idea how hard it will be to come up with a vaccine, or if it's reversible for those already affected. Or what other side effects there might be. For all we know, it could be in the process of mutating into a more virulent form which renders everyone infertile." He paused. "Unless there's more you want to tell me?"

"No. There isn't. So we need more information to decide."

"We haven't got time to—"

William raised a hand. "Whether we move to stop the virus as fast as humanly possible, or consider giving it a little time then stopping it, our immediate course of action is the same. We must understand it better and we must start working on a vaccine. Before we do either, however, we need to confirm the causation."

"Okay, but our fertility data's patchy, particularly here in the U.S. because we can't get the big IVF players to help. Can you get us access to Genenco's records?"

"I'm not directly involved in the IVF business, so it may take a little time but I'm pretty sure I can make it happen, as long as you and your family don't rattle the cage too much. But if we start lobbing accusations around, they'll go into defensive mode and we'll get nothing."

"Okay."

"Meanwhile, why don't you assume the fertility drop is real and try to isolate the vector. I'll start investigating vaccine development labs so we're able to move fast once we have a vector to target…assuming we want to."

"Okay. And we keep in touch about our thoughts on how to proceed."

"Yes, but Ayden? I know I don't have to spell this out, but I'm going to anyway. When you make enemies in the big end of town, the first thing they do is dig. Hunt for vulnerabilities. Now I know it was decades ago, but there were six people in Gardner's team who knew all about us and three of them are still alive today. There are also at least four people in addition to your parents who had very strong suspicions. Our secret isn't bulletproof. Please, tread very softly."

"Understood."

"Also, we need to be able to communicate safely online, which means we need to seriously upgrade your phone security. Can I get the firm I use to contact you?"

"Yes."

"Even then, can I suggest we agree to be circumspect in our language when we're on-grid?"

"Yes."

Wednesday, December 28th

Ayden hadn't visited the university's molecular biology research lab before. He had to admit, it was impressive. It was essentially a very large room with a wall of glass down the middle. On Ayden's side of the glass, it was a bit like the lobby of a basic hotel. Researchers, apparently idled in chairs, clustered around low tables. Most were absorbed in their specs, controlling their experiments virtually, some were chatting with colleagues and sipping from glasses or cups or bottles. It looked relaxed and smelled of stale coffee.

On the other side of the glass, gleaming white machines sat in perfect rows, inscrutably silent and immobile, occasionally tended by technicians wearing clean-room coveralls, acolytes tending their altars.

Ayden linked into the lab interface, and the machines were suddenly surrounded by brightly colored icons that blinked or spun or glowed. Clearly, they were in demand even over the winter break.

"There you are, Ayden." His mother gave him a peck on the cheek. "I'm over here." She led him to a couch in one corner. "You didn't have to bring me lunch."

"The cafeteria's closed, and Dad says you wouldn't eat if someone didn't bring you food. And I wanted a break anyway. I don't really know why you're here though. You can run the experiments from anywhere, and the techs don't let you in the clean room do they?"

"What? Never." She smiled. "I do find, though, that they get to my requests a little faster if I'm hanging around, watching."

Ayden produced two containers of salad from the bag he'd been carrying. "So, any progress?"

"Yes." Cate hesitated. "I think so. We're a long way from confirmation but…you know I cultured human cells and exposed them to Ps Novo?"

"Yes."

"Well, they became infected by one of the viruses from the bacteria so we have a potential vector."

Ayden shook his head. "Oh good, a virus carried by a GM bacteria, carried by insects. Has the virus been engineered, too?"

"No idea. Next step is to try to confirm the linkage with infertility. Tess is organizing blood and semen samples from all the males involved in the project."

"Yes, I know." Ayden smiled. "She's having a great time. Marches up to us in the most public spot she can find and asks loudly for our semen."

Cate chuckled. "I think I'm starting to like that girl." Her face fell serious again. "I've ordered some mice to expose to the virus. If they become infertile, it'll be a pretty clear piece of evidence but really we need to get permission to test primates, rhesus monkeys at least, if we can't get chimps. We're fast approaching the point where we can't make significant progress without serious resources."

"Yeah."

Thursday, December 29th

It was late, almost ten in the evening, and Ayden was exhausted, but he hadn't heard from William for days.

On the train back from visiting William in San Francisco, Ayden had watched closely as William's security consultant upgraded his phone. Zachary's avatar had pointy ears and wore some sort of uniform. After a few moments, Ayden remembered the vid series from his youth. Zachary also maintained a running dialogue of every patch he applied. He didn't draw breath. He explained that none of the freeware sites were worth a damn and that "cheapo" security subscriptions weren't worth much more. Over 20 percent of Ayden's apps were vulnerable to unpublished exploits, according to Zachary, and the firewall needed to be ditched completely.

His replacement firewall was modeled on military ware and would stop almost anything. He also explained that there was normally a two-month wait for his services but with Mr. Hanford as the client, Ayden got special treatment and Zachary got special rates.

After Zachary finished, Ayden tried to get back to work, only to be interrupted by an almost constant stream of alerts and clarifications from Zachary's avatar, thankfully less loquacious than the original, as the new firewall bedded itself in.

He'd finally, after a few days, reconfigured the alert agent to a minimal text-only form and the torrent of interruptions had slowed to a trickle.

And it meant Ayden could contact William online.

[Any joy re the IVF records?] Ayden sent.

William's reply was almost immediate. [Not yet. I've created a plausible but bland reason for requesting the data and promised to share any results but Rossouw, the guy heading up IVF, is being cagey. His hackles are up. Says a colleague told him there's a rumor doing the rounds at the University of Washington about falling IVF success rates. I presume that's your Tess.]

[That doesn't make sense. IVF can work around the capacitation problem.]

[Since when do rumors have to make sense? Don't suppose you can put a lid on her for a week or two? If I push Rossouw too hard, he'll demand to know everything.]

[Easier said than done.]

Friday, December 30th

Ayden sipped on a steaming cup of green tea and watched Tess sip a soup bowl–sized cappuccino and pick at a muffin. The café was a cozy haven of warmth after they'd crunched though a thin layer of fresh snow on the way from Tess's shared apartment.

"Do you remember about six months ago, I told you I met with that Genenco Director, William Hanford?"

"Vaguely. You convinced him to call off MTC's dogs after they nearly got you fired."

"We met again on Monday. He promised to try and get us all the data from Genenco-related IVF facilities. It'll give us huge coverage not only in the U.S. but also through Africa and India. But we need to keep a low profile for a while."

"Why? What's going on?"

Ayden took a deep breath. "Genenco has something to do with this. William admitted to me that the infected mosquitoes were a Genenco research project that got out of hand."

Tess's eyes went impossibly wide, then narrowed. "When did he admit it? Six months ago or Monday?"

Ayden hesitated.

"That's what you weren't telling me," she said. "You agreed to a cover-up in exchange for his help becoming science's golden boy." Tess's voice went flat, her face wooden.

"No, but I did agree it was more important to fix the problem than rake over old coals, and…yes, William helped us to get the bee intervention role."

She just stared at him.

"I know, I know," Ayden said. "It sounds like good old-fashioned corruption, but it's not. At least, I won't let it be. It's more along the lines of an alliance aimed at solving a problem. It's certainly not going to stop me finding out exactly what's going on. But if this thing blows up before we get the data, Genenco will go into ass-covering mode and it'll take six months in court to get anything out of them."

"When, Ayden?" she said. "When's he going to hand over the data? How long are we supposed to wait?"

"I don't know, a week. Maybe two at the outside. In Europe, we've got good fertility data but poor mosquito data, here it's the reverse. We really need Genenco's data. With it we can go to the EPA with proof rather than inconclusive evidence and speculation."

"Well, I'm not sure we can wait any longer. Jack Cooper wasn't an isolated incident. Nearly 40 percent of the male researchers who came into contact with infected mosquitoes have sperm antibodies."

"Okay, but do you really think a week or two will make a big difference?"

"People have a right to try to avoid infection. It's all very well for us right now. It's winter. But what about the tropics? What about the southern hemisphere? We have to go public."

"You're right, as tempted as I am to disagree," he said, with a wry smile. "But that's exactly why we need the data, and right now we have a one-off opportunity to get it quickly and easily."

"What do you mean, 'tempted to disagree?'"

Ayden tried to wave it off, but she held his gaze.

"Okay, let me run a hypothetical past you. Imagine this virus was the perfect male contraceptive. A hundred percent safe and effective and easily reversible by the individual. No messy medical procedure, he just has to pop a pill. Imagine a worldwide plebiscite. Would you vote to release the virus?"

"No more unplanned pregnancies, ever, anywhere? Sure."

"Even though the people who voted against it would also be infected?"

"Of course. That's the way democracy works. Laws restrict all sorts of individual rights at the will of the majority. Besides, in your perfect example, there's no downside. Any guy could make himself fertile again any time. Hell, I wouldn't just vote for it. I'd campaign for it."

"Let's say I've got a vial of this perfect stuff in my pocket. Do we campaign for a worldwide vote we know will never happen— or do I just open it?"

Tess locked her eyes on Ayden's. "Where are you going with this, Ayden?"

"Nowhere. But…look at the state of the world."

Tess looked around the cafe, as if trying to find an escape from the conversation. No one was paying any attention to the two of them huddled in the corner.

"My god! It's that damned Hanford, isn't it?" She whispered furiously at him. "Ayden? Is this thing really a lab accident, or is it some crazy attempt to control the population level?"

"William claims it's completely accidental."

"And you believe him?"

"I don't know."

"But he's trying to get you to…what? To just stand by and let it infect everyone? Do nothing?"

"No. We've agreed to an urgent effort to understand the phenomenon and, assuming it's the virus, to creating a vaccine as fast as possible, and then hopefully some way to reverse the damage in those already infected."

"But?"

"Well, the question he's posing is, if, before we stopped the virus, it applied a tiny bit of a brake to the population, would that be such a terrible thing?"

"Yes, it would. It—"

"Worse than the damage caused by all the extra people?"

"This isn't your hypothetical utopian virus we're dealing with here, Ayden. Do you have any idea what IVF involves, quite aside from a lot of money?"

"Vaguely. It's not a fun process, I know."

"It's an awful process! I watched my cousin, Nikki, go through it. Months of needles and tests and hormones that messed with her head and then a painful procedure to extract her eggs. After that, the embryo implantation was only uncomfortable and intrusive but then…her first baby miscarried. So she did it again, and the embryos didn't take. It took her nearly a year to summon the courage to try a third time."

Ayden didn't respond.

"What's he done to you?" Tess said. "How could you even consider delaying? What if we can't develop a vaccine quickly, or cure those already infected? It took forty years to beat HIV. What if the virus mutates and is too successful and it sterilizes everyone? A few months ago you would have been critical of anyone taking such risks with plants, but this would be a genetic engineering field trial with the whole of humanity."

Hands flashing, Ayden shut down his phone gear. He'd examined his new firewall closely and trusted that it would be very hard to hack through, but he couldn't be completely sure that Zachary wouldn't have given William a back door. He gestured silently for Tess to do the same. She did so, with a frown.

"What's this all about?" Tess said.

"Just being careful," he said quietly. "You're right, Tess, this isn't some hypothetical perfect solution. It is too risky. It certainly could be anyway. We just don't know enough. Look, I'm not actually saying we shouldn't try to understand and counter this thing as fast as possible. All I'm suggesting is, to get the data, it may be worth waiting a week or two before going public. Being strategic. That's all."

"Wait. Do you trust this guy or not?"

Ayden sighed. "I know he's not telling me everything. He's as much as admitted that. And I'm sure there are all sorts of things going on behind closed doors at Genenco. But where does William really stand? Is he running things or just caught up in the middle of them. I don't know. But I do know that he cares. He's worried about the mess we're making of the world, about where it might lead. And so am I."

Tess peered into his eyes. "What are you…" She drew back from him, shaking her head. "What if he keeps stalling?"

"Then we'll have learned something, won't we? But if you go public now, we won't get the data, and we won't know whose side William is on."

Tess stood, looked down at him. "Right now, I don't even know whose side you're on."

She grabbed her bag and headed for the coat rack.

Saturday, December 31st

[Hi Mom.]

[Hi Sweetheart. Tess went to George. They're on their way here now. I refused to discuss it online.]

[I'll be there in twenty.]

* * *

Tess was in full swing when Ayden got to the family condo.

"—any evidence, but I'll bet that damn Hanford's behind it."

"Calm down please, Tess," Cate said. "Hi, Ayden, darling. Tess, the situation is simple. We're scientists. We revise our lab and field protocols. We report our findings. We present the evidence." She accepted a kiss on the cheek from Ayden. "We don't go off on personal crusades against powerful and possibly innocent people on the basis of vague circumstantial evidence."

"Vague circumstantial evidence?" Tess said. "What do you want, a contaminated syringe under the pillow?"

"Actually, yes. If the powers that be think there's evidence against Genenco or Hanford, they can go after them. The world isn't a fair place, Tess. If people like us attack powerful multinationals then we better have a smoking gun. Otherwise, maybe anyway, they get off scot-free and we get squashed."

"She's right, Tess," Marc said. "There are mechanisms and people dedicated to pursuing blame and dispensing justice."

"Listen to an old war-horse, my dear," George Reyes said. "Think tactically. If we attack Genenco, they come back at us with everything they've got—attack our science, our credibility, us personally. They make it much harder for us to convince the authorities to throw real resources at this. If we don't attack Genenco, they have no reason to come after us. Even if you are right, and someone there deliberately cooked this virus up, they'll have to be circumspect about attacking us. On the face of it at least, Genenco will have to cooperate with the investigation."

"Okay, Dr. Reyes, but later, if we find more evidence…?"

"Then we reassess. I assure you, my dear, if I think we can prove that someone is deliberately attempting to manipulate the human race, I'll go after him. I'll throw this old body onto the sword, if necessary. Deal?"

"Yes, sir. Thank you," Tess said, eyeing Ayden sideways.

"But…" Reyes got her attention back immediately.

"But?"

"In the meantime, we limit ourselves to presenting the evidence, and we do it in writing. No speculation. No hearsay."

"Yes, sir. Dry, cautious scientific language only…until we find the smoking gun."

W.R.R. MUNRO

9 :
JANUARY 2034

The newsreader's Asiatic face was almost as animated as his voice. He was standing in picturesque leafy gardens with the impressive columns and blue dome of the Uzbeck parliament building in the background.

"In breaking news, the President of Kyrgyzstan has accused Uzbek security forces of being behind the repeated sabotage attempts against the Kambaratinsk Dam development. Further, he has put those he describes as anti-dam terrorists on notice that in the future they will be shot on sight. In a tit-for-tat exchange, President Taymazov of Uzbekistan claimed that his nation had nothing to do with what he described as 'Kyrgyz civil unrest.' The Uzbek President went on to say that the dam was in violation of international laws and would threaten the agricultural output of the Fergana Valley and that his nation would be lodging proceedings with the International Court of Justice."

EURASIA NEWS, "WATER TENSIONS ARE ON THE INCREASE AGAIN IN CENTRAL ASIA," EURASIANEWS.CENTRALASIA

Sunday, January 1st

[What happened?]

[I tried to persuade her, ended up doing more harm than good. She went straight to Reyes.]

[You know we've no chance of getting the IVF data now, don't you?]

[She would have gone to him sooner or later anyway.]

[Later would have been better.]
[Yeah.]

Tuesday, January 10th

A week later, George Reyes met Jim Reynolds at the Club. They'd reluctantly left the wing-backed chairs in the Fireside Lounge, made their way to the Heritage Dining Room and ordered lunch.

"I'm sorry, George" Jim Reynolds said. "You should have your hands full with the bee intervention anyway."

"Come on, Jim, that's horse manure and you know it. The intervention's running smoothly, and this is a thousand times more important. And it's our discovery. We're right in the middle of the whole thing. Why cut us out?"

"I'll say you're in the middle. Too far in." Jim slowly swirled his wine glass and his long face regarded George's somberly.

"We ask you to investigate a Pacific Northwest bee issue, and suddenly you've got people all over the world collecting mosquitoes," Jim said. "Then you come in telling us a tale about human fertility and hinting at a cover-up conspiracy involving the 'crème de la crème' of the scientific establishment. So we need some calm, uninvolved people to look at the data you've produced and tell us if the shadows you are jumping at are real."

"You've known me for nearly twenty years, Jim. Do I make a habit of running around screaming that the sky's falling?"

"No, George, you don't. That's the only reason we took it seriously. Actually, it's out of my hands now anyway."

Reyes looked sharply at Reynolds.

"Who?"

"The CDC."

"Ah."

"Which is why they're sending a team around to your lab this afternoon to collect all your samples."

"What?"

"I don't believe your lab is Biosafety Level 2–certified, George. So if you're right about this virus, then it's a federal offense for you to keep handling it."

"Don't be silly, Jim. We came to you with this. We've also tightened our protocols dramatically, but the real point is that it's out there anyway…it's all over the world."

"Agreed but—assuming the threat is real—you've also introduced it to unaffected areas, exposed people who might otherwise not have been exposed."

"Hmmm."

"So they have no choice but to shut down you down."

"Hmmm. Or bring us inside."

"It's in safe hands, George. You guys just keep doing your job, which is a problem with bee pollination, by the way, and if memory serves, there isn't any correlation with the infected bee spread is there? It's just with the mosquitoes?"

"Correct as always, Jim. You'll keep me informed?"

"Always George. When I know anything, you'll know."

"Thank you." George prodded at the fish on his plate, skeptically. I don't suppose there's any chance this really is sole, though it's hard to tell when it's overcooked and smothered in sauce. When is this Club going to get a decent chef?"

W.R.R. MUNRO

10:
FEBRUARY 2034

The tiny body convulsed weakly. A thin dribble of vomit tricked down her chin.

"Davi! Quickly! More water!" The infant's mother said. She thrust a towel toward the camera, and Davi's hand reached out to take it. The viewpoint bobbed as the boy walked down a narrow corridor into a grimy bathroom where his wizened grandfather was sitting on the floor, motionless, eyes shut.

"Avo?" the boy said, uncertainly.

The emaciated old man opened his eyes listlessly but smiled for the boy. "Davi. You are a good boy," he said in a scratchy whisper.

"You okay, Avo?" the boy asked.

"Fine, Davi. The tiles are cool."

The camera bobbed again, and Davi's hand reached out to turn the bath tap, but the spout remained dry. He leaned into the bath, wiped up the last little puddle of brown water and rushed back to his mother.

"Quickly, Davi," she said, then, "no, wet it properly. The fever is bad again. I need to cool her."

"There is no more water, Mama."

"Take the bucket, Davi. You must find some water."

"No one will give me water, Mama. I tried before. You remember."

"She'll die, Davi. You must find water. Sell your camera. Swap it for water."

"Yes mama."

The camera went dead.

HANYU, "SOUTHERN EUROPEAN WATER SHORTAGES REACH A CRISIS POINT. SHOULD WE BE DOING MORE TO HELP?," ANON.ENGLISH.HANYU

Friday, February 3rd

"Umm, Ayden? No offense, but I think you may've missed the point." Tess had popped in to Ayden's virtuality and was looking around at the blizzard of icons surrounding them.

"Huh. In what way?"

"Well, isn't the point of these virtualities to give you easy access to all your resources without the clutter?"

Ayden examined her. He could see she was making an effort to return to the easy banter they used to share. "Clutter? What clutter? I need this stuff."

"You need to do some serious housekeeping." Suddenly, her window morphed into a huge avatar threatening his scape with a massive broom.

Ayden moved fast, shrinking her to mouse size and pointing an eraser at her. "Drop it and back away slowly, lady."

Ayden watched her try something, but he'd already stripped her back to read-only access.

"You win. But only 'cause you've got admin rights."

He let her go, and in a flash she was back in a standard, frameless 2D window. "Seriously though, how do you work in this mess?"

"Actually, it's less chaotic than you think. Look closer. What's unusual?"

She shook her head.

Ayden cranked up the scape's color saturation. "And now?"

"Ahh, some sort of color coding?"

"Allows me to layer different projects in the same proximity. Bees in orange, Hunt River in blue, human fertility in green, et cetera." He backed the color saturation off. You can still see the colors now that you know what you're looking for, can't you?"

"I guess so. Look, I know you're juggling a lot of different stuff at the moment. I forget that sometimes and expect you to drop everything for my priorities and I'm sorry."

"It's okay, Tess."

"But it's really important, and I think maybe you need to take a look at your level of objectivity when it comes to Hanford, you know?"

"Uh-huh."

"Damn, that wasn't a very good apology was it?"

"Don't sweat it, Tess. It's been difficult all round. I'm not angry that you brought George in. You did what you thought was best. Hey! Wanna see something cool?" Without waiting for an answer, he pushed everything down and the entire scape flattened out to two dimensions at their feet.

"Icon carpet?" she said, peering.

Suddenly the "floor" receded so they soared high above it, and he cranked up the color saturation again.

"Whoa. Okay, now I see the pattern. Is it...fractal?"

"It may be, though there are gaps and it's not fully scalable. I suspect it would be if I had complete information." He restored the 3D scape then pushed sideways so it was now a 2D wall, which then receded in an instant.

"Vertical patterns as well."

"Any angle." He restored the 3D scape again. "Now you know what you're looking for, can you see the pattern in 3D?"

She peered at the scape for a while. "Not really."

"Takes practice."

"So, how much time have you spent arranging the icons to get this effect?"

"Arranging icons? Nada. Zip."

"No way."

"I just redefined some of the parameters for association space representation."

"Oh, is that all?"

"The pattern seems to emerge from the complexity. I'm gonna ask Dad to try to come up with a formal proof."

"That's...well, from anyone else it would be frankly unbelievable. Is it actually useful?"

"I can't prove anything yet, but I think it might be. When you're working through a puzzle, patterns are usually useful. At the moment, I'm looking at the gaps. If associational information really is fractal, then there shouldn't be gaps at any scale, so the gaps indicate a lack of information. In most cases, imperfect information doesn't matter too much, but some gaps are at important junctures."

"Uh okay, so what's it told you so far about human fertility?"

"Nothing concrete yet. Sorry. Not enough data."

"Well, I believe that. Ayden, I'm climbing the walls. A month without word from the CDC! It's unbelievable!"

"More like three weeks, but yeah it's frustrating. Problem is— hold on, Tess. Mom's inviting us to join a call. From George. Might be news."

The virtuality whisked away, and they were in a simple conference call scape. Blue sky, the odd fluffy white cloud, grass extending evenly to the horizon, simple frameless windows. Cate and Marc were together in one, George, Tess, and Ayden each had their own.

"I'm afraid the news isn't good," George said. "They haven't finalized their investigations but they say they don't see a statistically significant correlation."

"Hey! There's nothing wrong with my stats," Marc said.

"I'm sure you're right," George said. "But as we well know, our data was patchy."

"Of course it was. We didn't have the resources—" Tess said.

"Exactly, which is why we brought them in. They do have the resources, and you'll be interested to hear that the U.S. IVF companies, including Genenco, cooperated fully. They handed over all their IVF data, and it doesn't show significant correlation with infected mosquito populations."

"We need to see the data," Ayden said.

"Well, unfortunately, Ayden, they're under no obligation to hand it over. In fact, they're not allowed to. Commercial in confidence."

"Perfect. What about the direct causal evidence?" asked Cate. "What about our male researchers? What about testing rhesus monkeys, and if they test positive, chimps?"

"The mice were immune," George said.

"Last time I checked, mice weren't primates."

"They also tested pigs, which did become infected with the virus, but it had zero impact on fertility. They tell me there isn't enough evidence to justify primate testing. You know how hard it is to get primate approval, Cate."

"But that's to stop them from being treated inhumanely. This is just exposing them to a virus that the CDC thinks is no threat! That's ridiculous."

"It's also a reaction to all the bureaucratic red tape, lifestyle assurance, monitoring costs, and political fallout associated with primate testing."

"Can't we just piggy-back an existing study? There must be some experiments running that wouldn't be adversely affected by the test animals getting the flu for a few days."

"I'm sorry, my dear, but that sounds like someone from the private sector talking. In the public sector, one either justifies a study or one does not. Besides, you know how keen scientists are to introduce additional variables to their experiments once they finally get them approved and underway."

"Well, it's not important anyway. It's just the future of the human race we're talking about here," Marc said.

George just smiled sympathetically. "Clearly, the CDC didn't agree."

"The bastards." Tess said. Everyone looked at her. "They have the CDC wrapped around their fingers."

"No, Tess. Stop right there if you want me or anyone else to continue to take you seriously," George said. "Conspiracy stories won't wash, they'll just push you to the lunatic fringe. I don't believe the CDC would help cover up something like this. I know people there, and I'm telling you that they take their roles very seriously. If they say they don't see a significant link then I believe that they don't see one."

"Then Genenco messed with the IVF data, or…"

"You're grasping at straws, Tess," George said. "I looked at your data and I agreed there might be something there, but just *might* be." He looked around the room. "Now I have the greatest respect for each of you, but I also respect the CDC. They say there's nothing there. So, unless and until some more evidence shows up, that's good enough for me. Some of you may wish to follow up the fertility angle on your own time, but—and I mean this—as far as I'm concerned, our job is to focus on bee pollination levels, so let's do our jobs." George eyed Tess "And no more wild allegations, not in public, not to anyone outside this room. This project doesn't need any more blows to its credibility."

George Reyes thanked the team for their time, and his window disappeared.

"George thinks we've embarrassed him," Marc said.

"We have," Cate said. "The CDC investigated and didn't find a correlation. Mice and pigs were unaffected. Were we wrong?" Cate asked, looking first at Marc then at Ayden.

"Our science was good, as far as it went," Ayden said.

"But…?"

"But we didn't—and still don't—know enough. We jumped the gun."

"What are you saying, Ayden?" Marc said.

"I'm saying we need to keep going. More research."

They all just looked at him.

"I don't know. I really don't. But I will find out."

Tuesday, February 7th

The narrow, poorly lit room in Marc's specs was dominated by a large blackboard. He surged forward, brandishing a white piece of chalk and a dusty felt eraser. He circled his probability density function, scrawled "df=?" and backed away, only vaguely aware of the dark, waxed wood and scuffed leather furniture. He was a long way off but the approach felt right. Just possibly he wasn't too old to derive a significant proof.

"Marc," Cate called from over his shoulder somewhere. "Marc?"

With a gesture, he cleared his specs as Cate rushed into the living room.

"Oh, sorry. Busy? I think I've found someone who can help us," she said.

"Don't tell me, another admirer?"

"You behave, Marc Walker. You will be relieved to note that Dr. Ramy Da Silva doesn't have designs on me. You're more likely to appeal to her. We haven't met, but we know each other's work. Ramy is with the Republic Polytechnic in Singapore. She's doing a behavioral study on a group of chimpanzees and is confident that mild flu symptoms wouldn't cause any difficulties. She hasn't committed to anything yet, but she's happy to discuss it."

"Sounds like fun," Marc said. "But how do we get approval to ship infected foreign mosquitoes into Singapore?"

"We don't try. Some of our data came from Malaysia, so there are infected mosquitoes there. I'm going to fly to Singapore and talk to Ramy then, hopefully, drive into Malaysia with collection kits. I'm not going to let anything foul this experiment up. We may not get another opportunity."

"What? Who's going to cook and clean up around here and darn my socks?"

"I thought I told you to behave, old man."

"Seriously though, I'm not letting you drive all over Malaysia by yourself. I'm coming too."

"Sorry, darling, you have to stay here and hold the fort. Besides, Ramy has offered me a couple of undergraduate students. I suspect I'll be well looked after."

"I'm starting to like this Dr. Da Silva."

"Me, too. Book me a flight? I'm going to start packing."

Thursday, February 9th

Cate spent the flight working with her phone, dealing with the myriad small details associated with the ongoing bee intervention. They figured George Reyes would turn a blind eye to their activities only if the bee project proceeded smoothly.

When Cate finally cleared customs with her baggage, she was surprised to find Dr. Ramy Da Silva herself waiting, looking identical to the photo Cate had displayed in a window of her specs. She was tiny and immaculate, wearing a striking purple pantsuit. Her glossy black hair and ageless, unlined face made Cate feel ungainly and old, but her smile was warm and welcoming.

When the terminal doors opened, Cate was shocked by the wave of heat and humidity, but they were soon in Ramy's university car.

Once they got going, Ramy turned to Cate. "There are so few places where one can talk completely freely these days," she said.

"Umm...yes," Cate said.

"I'm talking to you partly because George Reyes is an old friend of mine, and he asked me to."

"He did? I didn't even tell him what I was doing."

"I checked up on you. I messaged him. Sorry."

"Don't be. I'd have done the same."

"His response was vintage George Reyes' cloak and dagger."

"Cloak and dagger?" Now Cate was really puzzled.

"Yes. I'll read it to you." Ramy ignored the road as the car navigated its way through Singapore's streets. "His response was, 'Cate Walker is a very dear friend whose judgment I hold in the highest esteem and to whom I would deny nothing. I would not hesitate to support her any and every endeavor."

"That's it?"

"Yes."

"That's lovely but...vague." Cate thought it awkwardly formal but didn't want to say so—perhaps he always corresponded with Ramy in that tone.

"It's George being careful, sending me a strong message but avoiding any specifics, maintaining a clear distance from your request itself."

"I'm not surprised he didn't want to talk about it."

"Which brings us to the other reason I agreed to talk to you. Curiosity. You have an EPA mandate, but are clearly going far beyond it. You're avoiding official channels when asking for assistance, and George is in cloak and dagger mode. So I just have to know…what's going on?"

Cate told Ramy everything about the bees, mosquitoes, and capacitation. Rather than going to Cate's hotel or to the university, they went to Ramy's apartment "for tea."

The apartment was small but exuded a calm peacefulness that entranced Cate from the moment she arrived. When they entered, Ramy struck a gong and paused with her eyes closed then offered Cate the little padded mallet with a smile. Cate hesitated, surprised and a little uncomfortable with the shared intimacy of the moment, then followed suit and found herself relaxing as the warm, rich sound washed over her.

The furniture was mostly dark wood, and an exquisitely carved and painted frieze dominated the longest wall of the little living area. It depicted a festival or procession focused on some sort of mythical winged creature. There were bright, colorful cushions everywhere, and several palm plants with their red stems flowing into up-thrust green fronds.

After they had talked for a while, Ramy insisted that Cate cancel her hotel booking and stay in her spare room. By then it was dinnertime, and Ramy cooked a simple but delicious vegetable curry. During the meal, Ramy refused to discuss anything related to their work. Instead, they spoke of family. After Cate helped Ramy handwash the dishes (ignoring the dishwasher she could see under the counter), they sat down with small cups of delicious tea. Ramy took a tiny sip and began.

"If there is even a tiny possibility that your suspicions are correct, then it must be fully investigated."

"Unfortunately, the CDC thinks they've already done that."

Ramy smiled. "No, my dear, it must be investigated to your satisfaction."

"Mine, in particular?"

"Yes."

"Why. What's special about my opinion?"

"Do you have any idea how many people George Reyes describes in the terms he used for you?"

"No."

"Three people that I know of...and I've known George since college days."

"Oh."

"So tell me what you need to hunt mosquitoes."

"Thank you for agreeing to help me, Ramy. Simple carbon dioxide traps work well, so all I really need are small carbon dioxide bottles, buckets, mosquito netting, and some wire and tape."

"You're going to build the traps yourself? Wonderful!"

"Ideally, I'd like to build about a dozen traps, spend a day in Johor setting them, then pick them up the next day. We know there are infected mosquitoes in the Muar area so I thought I should go straight there."

"Easy. One of my undergraduates is my cousin's daughter, Salma. Intelligent and very trustworthy. Her boyfriend is a physics major, but he'll do anything for her. Funny-looking boy, she calls him Elvis, I don't know what his real name is. Anyway, they will be here in the morning with everything you need. You can do your construction here, then take my university car into Malaysia. You may even get your traps set tomorrow. You'll stay with my brother and his wife in Melaka Tenga, which is very close by, and with luck you'll be back in time for dinner on Saturday with a boot load of bloodsucking little plague carriers."

"Wow. That sounds easy."

"What do you need to test your mosquitoes to see if they're infected?"

"I brought some lab-on-a-chip kits with me, which we use to detect the bacteria. Now we know that's not the infertility vector, but it's proven to be a reliable indicator."

"So, assuming you have infected mosquitoes, we then expose my chimps to them."

"Well, some of them…you said you were studying forty-six animals?"

"Yes, including fourteen sexually mature males."

"Would it be possible to separate them into two isolated groups for a few days? Keep one of the groups uninfected as a control?"

"Yes, I think we can do that. We've collected two samples from each of the males, which I've had frozen just in case. We saw no evidence of sperm antibodies at all."

"So it looks like we're starting with an uninfected population. Excellent."

"Now, Muar is quite conservative. You would probably find it easier to get the job done wearing a headscarf. I have some beautiful scarves in a lovely light cotton if you'd like to try a couple on."

Sunday, February 12th

It took longer than expected to set out all the traps. Cate didn't get back to Singapore until mid-morning Sunday. She went directly to the zoo where Ramy was supervising the separation of the chimpanzees and followed her specs' prompts to find the Congo River Habitat.

The zoo was in the middle of a natural rainforest in Singapore's central catchment area. Cate carried a bulky but lightweight bag over her shoulder, inside which were plastic sample jars full of mosquitoes. Heading down the Rainforest Walk, Cate felt as if she could be in the middle of a jungle anywhere in Malaysia or Indonesia. The landscape was overwhelmingly lush and used concealed moat barriers to separate the wildlife from the people. Hoots and hollering followed her. 'Siamangs,' according to her specs. 'Tamarins' chirped and squeaked, and a dazzling variety of butterflies fluttered silently though the shadowy green world beneath the canopy.

"There you are," Ramy said when she saw Cate approach. Her long shirt was filthy, and her face glowed.

"Hi, Ramy. Everything okay?" Cate said.

"Oh, yes. Perfect. Well, almost. We separate the chimps into sub-groups all the time, but usually we let them decide who goes where. Separating even numbers of sexually mature males was…a challenge. Right now, we've got eight in one group and six in the other. Is that close enough to half, do you think?"

"Yes, that's fine."

"Shall we make the smaller group the control, then?" Ramy said.

"Yes, please. I ran lab-on-a-chip analyses through the night. The mozzies test positive for the bacteria. We'll do more comprehensive testing to confirm viral DNA back in the States, but it looks like we're good to go."

"Well, the test kit reads negative for the mosquitoes we caught here, so it looks like we've got uninfected insects for the control chimps."

Ramy led Cate past the front of the chimpanzee exhibit, a dense jungle setting full of trees and logs for the animals to clamber over. At the back of the exhibit, a small waterfall cascaded down a rock face. To one side was an artificial cave with a large viewing window set into it. A sign provided instructions to visitors explaining how they could use their phones to access the twenty micro-cams dotted about the exhibit to get an even closer view.

There was a door in the back of the cave, which clicked unlocked as Ramy approached it.

"Thankfully, there are four separate sleeping caves back here," Ramy said as they entered a long, narrow room. Aside from a long table and benches, it contained four more doors.

A musky, damp locker-room smell pervaded the air. The exposure boxes were waiting for them on the table, transparent cubes with round sliding hatches on one side. Half were already labeled and full of mosquitoes. Cate and Ramy busied themselves trans-ferring Cate's infected mosquitoes to the other exposure boxes and labeling them.

"There's no hurry," Ramy said. "The handlers need a bit more time to get them settled anyway." She gestured and Cate flicked the offered link.

A window opened in Cate's specs showing a young woman in zoo uniform bending over a sleeping chimpanzee, shaving a little patch of forearm free of fur.

"You've sedated them?" Cate said.

"They're not unconscious, just sleeping," Ramy said. "I think we should wait until they're finished shaving before we go in, to minimize the disturbance."

A little later, they quietly entered the door with the handwrit-ten "control" label taped to it. Cate followed Ramy, carrying the exposure boxes. Inside, the musky smell became pungent.

Ramy introduced Cate to the handler, Marie, sub-vocally, and they got to work. Taking each animal in turn, Marie would gen-tly hold the chimpanzee's arm out, and Ramy would hold the exposure box against the shaved patch of skin and slide the hatch aside, allowing the mosquitoes a brief feast. Cate captured a video log with her specs and, hands flicking, used a university record-keeping app to keep track of which sample box was used for each chimpanzee, making sure the box radio frequency tags matched the handwritten labels and were associated with the impressively complete suite of biometrics being captured for each animal.

[Great data,] Cate subvocalized. [Is all this for our experiment?]

[No,] Ramy subvocalized. [The zoo collects this data all the time as part of their animal welfare program. See here?] She indi-cated a patch of suspiciously clean, shiny black fur on the chim-panzees shoulder. Cate peered closely.

[It's fake fur with an imbedded sensor cluster glued to the skin.] Ramy subvocalized. [Here too.] She pointed to another patch of fake fur over the heart.

Thursday, February 16th

Within twenty-four hours, the infected chimpanzees had shown signs of fighting a mild infection, and two days later, their biometrics showed them to be clear.

A day of hilarity followed as Cate learned more than she ever wanted to know about collecting sperm samples from chimpanzees. This involved an artificial chimpanzee rump and some coaxing from the handlers. One chimp, which they'd named Errol, refused to have anything to do with the fake rump and had to be manually assisted, much to his obvious satisfaction.

Friday, February 17th

Ayden insisted that they not use any electronic means to discuss their findings so Cate had to restrain herself from messaging or phoning Marc or Ayden with the news. She swore Ramy to secrecy and headed back to the United States. At the airport, she gave Ramy a big hug.

"Thank you, Ramy. In a few short days, I feel we've become good friends."

"I also. You know, my curiosity wasn't all professional."

"What do you mean?"

"George and I were lovers at university. We've been occasional lovers since." Cate was taken aback with Ramy's frankness. Ramy continued. "I assumed you were his lover also, but you never have been, have you?"

"George and I became friends before I married Marc, but we were never lovers."

"Marc sounds like a wonderful man. You are very fortunate."

"Yes, I am. Thank you again, Ramy—for everything."

"And I believe George is right about you. Let me know if there's anything more I can do."

• • •

On the flight back, Cate received a message from George Reyes.

[My dearest Cate. I have finally tracked down some of that Portuguese morel I was telling you about. Now, if I recall correctly, this means you owe me a family dinner of your homemade pasta. Of course, the last thing I wish to do is apply time pressure, but my fungi vendor took pains to stress that the mushrooms should be consumed fresh to do them justice.]

Cate knew that George had no love for mushrooms. Certainly she had never discussed cooking mushroom pasta for him. More cloak and dagger.

[Back tomorrow. Dinner tomorrow night?] she sent.

• • •

Cynthia Chan's slim form froze halfway through snake-creeps-down.

[Stop,] she subvocalized even though she was alone. She sank into a comfortable half-lotus. [Rewind 1 minute and play.] She sat, immobile on the mat in the middle of her tiny but very private, leafy courtyard.

Cynthia had long since decided that the key to good surveillance was to be very fast at "getting to know" her targets. It looked like that approach may be paying off again. She listened carefully. It was innocuous except Cynthia remembered that Reyes didn't like mushrooms. She ran a quick search against the Reyes transcript database. There it was, Reyes carrying on at the Harbor Club when his lunch arrived with a mushroom sauce.

"Why spoil a perfectly good dish with fungus?" Reyes had asked.

Amazing how "useless" information could—just occasionally—be so valuable. Returning to her tai chi, she recorded a quick summary, appended the transcripts and fired it all up the line.

Saturday, February 18th

Cate updated Marc and Ayden over a salad lunch. They perched on high stools around the kitchen bench.

"There's no question. The mosquitoes caused the same sperm antibody problem in the chimps. In fact, the results were far stronger. Every male chimp exposed to the infected mosquitoes is effectively infertile without medical intervention."

"Against what, 40 percent in humans?" Marc said. "Wow."

"The experiment also constitutes strong evidence that the virus is the vector. The chimps all showed flu-like symptoms and a very strong immune system response," Cate said.

"Evidence but not proof. We need lab work," Ayden said.

"Maybe, but I seem to recall that the CDC shut us down."

"That was before they decided the infected mosquitoes were harmless."

Cate smiled for the first time that morning. "Indeed it was, my darling. Let's do it. I'm sure we can slip it under the radar at the Project."

"Hmmm. Maybe," Ayden said.

• • •

They were still trying to decide what to do when George arrived for dinner with mushrooms.

"I'm throwing these in the bin right now unless you promise not to feed them to me," he said.

Marc cooked them a cheese risotto. Ayden made a spinach and bean salad with the expensive roasted pine nuts that Cate had spent the afternoon hunting down. Cate updated George over a drink.

"Delicious risotto, thank you all," George said as they started to eat.

They sat around the small circular dining table down at the kitchen end of the living room. Cate had turned off most of the lights and lit candles, and their warm, flickering glow, together with her leafy plants, produced a cozy atmosphere.

"And special thanks for leaving out the mushroom. Now, as to why I inveigled myself this invitation. As you know, I've been pushing the CDC to hand over their data, but they would probably have left me dangling for months just to make a point. Fortunately, Jim Reynolds at the EPA also wanted to see it."

"He give you access?" Ayden said.

"Yes, lad, the old fashioned way." He handed Ayden a tiny memory card. "But please bear in mind that I'm breaching confidentiality by giving it to you and my career's fragile enough as it is, at the moment. I have to say I'm more relaxed handing it over in the absence of the lovely young Tess. She's intelligent, honorable, and charming but often winning the long game requires more patience than she's displayed so far.

"No problem."

"Why, George?" Cate said. "Don't get me wrong I'm—we're all—tremendously grateful, but why take the risk?"

"Because I had a good look at that data." George indicated the memory card lying on the table in front of Ayden. "Giving credit where it's due, the CDC did a thorough job. Not only did they get good coverage of sperm antibody data, they also collected mosquitoes where our data was thin or nonexistent. And based on that data, their conclusions are rock solid. There's no correlation."

There was a long pause.

"From which I'm forced to choose one of two conclusions. Either you three got it completely and utterly wrong," he held up a hand to forestall protest, "something I wasn't inclined to believe even before you told me the results of your chimp test, or someone is playing an illegal, complex, and dangerous game. That in itself doesn't surprise me overly. I've watched people do some crazy things over the years, but I can't for the life of me work out how they could have done it."

"I can," Ayden said with quiet conviction. "Assume someone within Genenco is behind it. They could easily have cooked their IVF data before handing it over."

"Of course, but then in those areas where their competitors dominate, the CDC would see a difference in the data sets," Marc said. "It would be like a flashing red light."

"Unless, focusing on those areas, they messed with the mosquitoes," Ayden said. Just think, Rhino Mozzies, desiccation-resistant eggs, engineered to quickly establish viable populations. Our indicator of 'infection' has been the bacteria, but we know the bacteria isn't the actual vector for human infertility. We're now pretty sure it's a virus, which is harder and more expensive to detect. I bet the CDC tested for the bacteria rather than the virus, just as we did."

"So all they had to do was breed mozzies with the bacteria but without the virus and spread them around!" Marc said.

"But you've no evidence they did anything of the sort." George said.

"No. What worries me is I'm not sure they'd have had time to breed and seed a clean population before the CDC started collecting samples. More likely, they had a supply of eggs stockpiled for the purpose." Ayden stopped, looked around. They were all waiting for him to say it. "Which implies extensive planning."

"It certainly suggests a much more active role than simply turning a blind eye to a lab error," George said.

"Yep."

"So, any ideas about who 'they' might be?"

"Plenty. But not a scrap of evidence," Ayden said.

"Care to share?" George said.

"Not yet."

Ayden and George locked eyes, and silence descended again.

Finally, Cate sighed and turned to Ayden. "So what's next?"

Marc simply looked at Ayden.

Ayden's universe pivoted. He had always been given a voice in the family's affairs, but the final decision had been his parents'. But now, Cate and Marc were looking to him for leadership. Ayden realized he hadn't really given them any choice. He'd been forcing his approach on them since they found out about the mosquitoes. Now, they were standing back and looking to Ayden to provide direction.

The scary thing was that Ayden knew they were doing the right thing. They were even more out of their depth than he was. Suddenly, a comfortable blanket had been removed.

"Hmmm, before you answer, Ayden," George said. "I think I should make myself absent. I suspect I can best serve our common goal by retaining my semi-insider status with the administration for the moment, so it's probably best if I don't hear what you decide. When the time comes, however, let me know."

The next few minutes were absorbed by the civilities of George thanking Cate for dinner and making his departure.

"Did I just hear right?" Marc said when George had left. "Did he just tell us that he'd be our man on the inside and sacrifice his career if need be?"

"Yes," Cate said.

"Why? Why would he do that for us?"

"Not for us. For what he believes in. To do the right thing."

Marc considered that for a moment. "Okay, so now what," he said.

Again, Cate and Marc looked at Ayden and waited.

"We get our own mosquito samples from the low sperm antibody areas. See if they're carrying the virus. We do the lab work. Confirm the human immune system reacts to the virus. Assemble the evidence."

"Then?" Marc said.

"I'm not really sure yet. Look, we just don't know enough. Obviously, we need to understand the virus better if we're to convince the CDC to look again, but it's also our only lead to finding out the 'why' and the 'who.' One more thing. Anything we do in a project lab, we have to assume they'll know about it immediately. Anything we do with a connected device, we must assume they'll learn about."

"You're saying we need to do this completely offline?" Marc said, as if he were being asked to work blindfolded. "In a completely isolated lab? And how are we going to fund that?"

Ayden looked at his parents.

"Oh, no. No, no, Ayden. We can't afford to bankroll a stand-alone lab."

"Yes, we can, Dad."

"That's a lifetime of savings you're talking about. It's how we'll eat when the jobs stop coming in."

"I know, Dad. Call it a loan. I'll pay you back."

"Surely we could take this to Reyes?" Marc turned to Cate. "See if he could get us a bit of funding on the sly."

"I couldn't ask him to dig himself in any deeper, Marc. Not until we have more evidence," she said.

He turned to Ayden. "What about William?"

"Not when we don't know where he stands."

"I'm sorry Ayden, we can't—"

Cate put a hand on Marc's forearm. "Yes, Marc, my love, we can."

She looked intently at Ayden. "We'll do it. Anything you ask."

"Hey, don't I get any say in this?" asked Marc.

"Of course you do, my darling. And as your faithful wife, I'll abide by your every decision. Which is why you're going to decide to do this." There was a long silence, Ayden and Cate looking at Marc. "We made our bed a long time ago, Marc. Now we have no choice but to lie in it." There was another pause. "We owe it to our son."

Marc let out a long sigh. "Okay, okay, damn it. But we do it on the cheap. No special purpose lab. We find a room somewhere we can lock up. We beg, borrow, and steal what we absolutely need."

Marc washed up, a process which seemed to involve a lot more banging of pots and pans than usual.

Tuesday, February 21st

They had a stroke of luck. A friend of Professor Sherman's heard about the receivership of a biotech start-up with a lab in Everett. The legal web would take many months, possibly years to unravel. In the meantime, the receiver couldn't sell the assets and welcomed the rental income. When they took possession, they discovered that much of the basic lab equipment they needed was there.

Unfortunately, their phones were useless.

"It's a simple choice, Dad," Ayden explained. "We either spend serious money on a top-flight security consultant, or we stay off grid and don't use anything wireless."

It was frustrating because, over the grid, their phones could access processing power that would put a turn-of-the-century super-computer to shame. They settled on leasing a server and an inter-face switch, which allowed them to connect several touch-screens. Servers were the behind-the-scenes workhorses that backed up the distributed processing power of the grid. They usually sat in large, frigid clean-rooms parallel processing together in their thousands. Connected together with cables.

"We can, at least, use them locally, just to access our server."

"No, we can't. What'd stop someone from parking outside the building and hacking in wirelessly?"

"Okay, okay, well…" Marc looked at his phone gear, the neck-band, cuffs, specs, and earpiece. "These things use skin conductance, don't they, to talk to each other, 'cause it uses less power than wireless transmission? What if we shut down their wireless functionality, disable the drivers completely? Then we just need to tape a wire to our wrists or something, and we could still use gestures and our specs to work."

Ayden laughed. It was like trying to take a security blanket from a toddler. "The switch doesn't have a skin conductance inter-face, Dad. And if we could rig something up, we wouldn't be able to use our specs outside the lab in case something got cached. Our phones need to be fully shut down inside the lab. If we really need to call someone or access the grid, we go outside so even if we're hacked, they won't be able to snoop in here. Even then, we still need to be careful—our information requests could be as revealing as a spycam in the ceiling."

"So…complete paranoia's the order of the day," Marc said.

"Yep," Ayden said. "They'll find out what we're doing sooner or later…but the later the better."

They ate, they slept, and they worked.

Tuesday, February 28th

Dwyer tiredly ran a thin hand over his bald head. He lay on the chaise lounge that looked out the big picture window of his forty-second-floor apartment. As usual, there wasn't a single light on inside the apartment, and Dwyer wore only a black dressing gown. He liked the drama of the city lights. And he liked to be invisible.

He had to admit he was impressed. These people were complete amateurs, but for over a week he hadn't been able to breach their security. They did all their work with their phones shut down, in a lab without windows, on a completely isolated network. They never discussed their work over the grid. Their precautions must have seriously compromised their productivity, but in a world where 99 percent of surveillance was electronic, they were effective.

Unfortunately, his employer, whoever he or she may be, was getting impatient.

Luckily, his targets were a bit more relaxed at home but even there, Dwyer had been forced to resort to an old, all but forgotten technique. It had cost him plenty, too. High-resolution video captured by a ridiculously expensive sniper camera aimed through the windows of the Walker condominium. Processor-intensive reflection filtering. An outrageously expensive specialist located somewhere in the U.K., judging from the guy's IP address.

Finally, however, it looked like it was paying off. Dwyer read through the lip-reader's transcript again;

{Woman walks to the window.}

"…question. The evidence is really strong, the virus is the cause."

{Pause. Response from out-of-sight range.}

"None. We've isolated the virus's DNA from Ramy's chimp cells and the lab-on-a-chip tests show a parallel immune system response."

{Man walks into view.}

"That's fabulous news. Vindication."

{Man and woman hug.}

"And we've confirmed that there are two populations of mosquitoes, both infected with the bacteria but only one with the virus."

{Man's response partly obscured then he stands back from the hug}

"…to go to the authorities yet?"

"I don't know. Ideally, we should access the CDC samples and prove that the high-fertility areas are virus-free… but I don't see us getting the access. But they can't turn away from this evidence! Let's talk to…"

{Woman turns away from window, then man also turns away. Readable footage ends.}

It was a short tidbit, but years in the game had given him a feel for when he had something worthwhile. With a flick, he paused Rachmaninoff's Hungarian Rhapsody and messaged his contact.

[Dwyer here. Have something for you. I'll do the drop first thing in the morning.]

W.R.R. MUNRO

11:
MARCH 2034

*T*win fences stretched across the landscape, each more than twice the height of a person. The first was sheathed in razor wire, but that wasn't stopping them. Thin, dry skin stretched over bones, the men and women at the front of the mob grasped the steel teeth and shook the fence back and forth, blood dripping from their hands to the bare, dusty earth below.

From behind them, a boy pulled from his mother's grip and joined them, offering his blood to the fence. On the other side of the pair of fences, young men in ragged army uniforms watched silently.

A fence post began to sway. A soldier with sergeant's stripes on his sleeves shouted at the mob, which ignored him. The fence swayed in ever increasing arcs. The sergeant shouted again, this time at his troops. They raised their rifles and opened fire over the heads of the mob as the post gave up its battle and the fence became a ramp. The men in front hesitated, but the mob behind them surged forward, pushing them ahead or climbing over the top of them. Another volley of fire from the soldiers achieved nothing.

Suddenly, a painfully thin and very young soldier rushed forward, throwing his rifle to the ground and stopping just before the second fence. He closed his eyes and touched the back of his hand against it. He was thrown backwards and lay moving feebly on the ground. Those at the forefront of the mob pushed back desperately, but the press of humanity behind them was relentless.

As one, scores of skinny bodies were pressed against the fence, their bodies jerking and starting to smoke before the electric current failed. The mob poured over the bodies and launched itself at the remaining barrier.

BioWatch vLog, "India Pakistan Border Tragedy as a 5 Percent Reduction in Food-Growing Season Exacerbates Food Shortages," biowatch.virt

Thursday, March 2nd

It was George's first and only visit to their makeshift lab. It was actually quite spacious, obviously designed for a dozen or more people. The center of the room was dominated by a cluster of small lab benches, each dedicated to an instrument. This appeared to be Cate's domain. The DNA sequencer hummed quietly. George suspected she ran it day and night. The centrifuge was spinning its test tubes like a miniature merry-go-round. One small bench had been cleared of equipment, and on it, rack-less, sat a server and switch sprouting cables. George hadn't seen so many cables for decades. Blue spaghetti cascaded from the table and snaked across the floor, connecting lab equipment and touch screens all over the room.

Marc had established himself on what had evidently been a meeting table. For a desk, it was huge but Marc had managed to cover its entire surface. Paper printouts showing endless tables of numbers and charts competed with anonymous junk to surround his screen.

Ayden's much more modestly-sized desk was empty, except for his touchpad and screen. George smiled. Even in a lab, Ayden left a small footprint. It would have been easy to overlook the fact that there was a third person working in the room.

Cate led George to a comfortable corner of the room. A very designer-looking couch and a couple of chairs made of fine, metal mesh surrounded a large, silver coffee table. There was a little kitchenette nook in the corner. Marc bustled about cleaning up cups, and Ayden put the kettle on.

Cate looked excited. "Actually we were just going to call you. We've made some real progress, and we need your advice. We've—"

George raised his hand, expression grim. "I'm sorry, Cate. I'm afraid I've got more bad news. I'm just going to come out and say this…you're off the project…both of you. They haven't dumped me yet, but they made it clear that I'm on the thinnest of ice." George waited for Marc and Ayden to sit down.

"Why, George?" Cate said.

"Unauthorized and unethical chimpanzee testing. That's what they're calling it."

"But…we only tested them with something that the CDC had declared safe!"

"For humans. They didn't make any findings regarding possible threats to chimps, half of which are now infertile. The Director of the Republic Polytechnic is furious."

"That's a bit of a fine distinction," Marc said then he stopped.

Cate's eyes had gone wide. "Oh no," she said. "Is Ramy…"

"Very definitely fired, yes," George said. "They've kept it quiet for now, but you know what a fuss this will cause when the press hears about it. Now that the hominid treaty has been internationally ratified, it'll be a circus."

"That's it then," Marc said. "Careers over."

"It may well be time for me to engineer a graceful retirement, too," George said.

"Please don't do that, George," Cate said. "We knew the risk we were taking."

"The damage is done," Marc said. "In any case, we now have evidence they can't ignore, evidence which will ultimately vindicate us. Actually, the chimp publicity could be useful. First we—"

Ayden abruptly stood. "Back in a minute."

* * *

[Someone is attacking my family.]

[I know.]

[Can't you stop them?]

[How? I'm not even sure who "they" are yet.]

[You're not powerless.]

[I've got some influence, but I'm not omnipotent.]

[You must be able to do something.]

[Not while you guys are on the attack. Speaking of which, I thought we were going to decide what to do together?]

[We were doing as agreed. Researching the virus.]

[Well, illegal chimp testing would certainly have raised a few flags but you must have done something more. Did you discuss going to the authorities? Insecurely?]

[We've endlessly discussed every course of action, William. We thought we were being careful.]

[Well, you've certainly got someone concerned. Look, it's a simple situation. Genenco's got a vigorous immune system. We just need to remove the threat. If you guys back off, then I can get Genenco to back off. Probably.]

● ● ●

[Hi, Ayden.]

[Hey, Tess. Good 3D. Where are you?]

[In a pokey little motel on the outskirts of Portland. You know the type, no cat swinging. Cardboard walls. They do boast a VR conference room though, so I thought I'd check it out. It's basically a cupboard with a few cams.]

[Portland? You're on the home stretch?]

[Eight or ten days. How'd Cate's trip go?]

[Definitely got a result. Not really something to chat about online. Feel like a home-cooked meal when you're done?]

[Sure. The exotic attraction of cheap motels wears thin, eventually.]

[Excellent. I'll give you a full rundown. Now, I don't suppose you know any propeller heads you trust? Someone happy to tiptoe down the line?]

[There is a guy. Name's—]

[No, don't tell me.]

[Okaaay. Let's call him…Dizzy. Anyway, he's a bit of a wiz. What do you need?]

[I need to find out who's behind an online identity.]

[I doubt that'd challenge him. Back when I lived on campus, I had a pretty scary online stalker. The cops tracked him as far as a botnet in Venezuela and gave up. Dizzy not only found him inside three days, he set him up. Made it look like the guy was operating a drop-zone for hacked grid passwords. Turned out the freak lived on campus. I heard the sirens when the cops picked him up.]

[Can Dizzy be trusted to be discreet?]

[There's the problem. He's a good guy but fundamentally insecure. He'd try to keep his mouth shut but he'd probably brag about it eventually.]

[Hmmm, we may have to live with that. You know this guy in the flesh though, right? You know who he really is?]

[Oh yeah. I introduced him to his girlfriend.]

[You remember that data miner I used? I showed you his out-there avatar?]

[Sure do. I'll get Dizzy on it.]

Wednesday, March 8th

"I'm sorry, but they wouldn't listen to me," George said.

With Cate and Ayden dropped from the bee implementation project and the family relying on Marc's freelance income, they'd moved out of the lab and out of the rented Seattle condominium. The Walker family had gone home.

For a couple of days, they'd just settled back into the comforting familiarity of their traditional English cottage in Mill Valley, California. Windows were unshuttered, potted plants were brought inside, and the house was given a thorough spring clean, all to the sound of Marc's extensive jazz collection, which he insisted had to be heard through huge, old-fashioned loud speakers in order to be properly appreciated. Cate decided she'd get the old vegetable garden up and running again, claiming there was nothing like homegrown produce and never admitting it was about the money. She'd started asking around the neighborhood to see if anyone knew where she could get some chickens.

They set up their leased server and got back to work. They had compiled their evidence, while Tess finished her traverse of the nation, tramping on and off trains with her field kit, collecting and testing mosquitoes.

Finally, they'd judged they had enough evidence, and George arranged another meeting with Jim Reynolds of the EPA.

Then Sanscor, Inc. dropped Marc halfway through a job and potential clients stopped returning his messages.

Now they sat in the living room, which Cate had furnished with an eclectic mix of old and new pieces to achieve a timeless atmosphere. Ayden knew it would be more bad news, the moment George walked in the door.

"They refused to even look. You've been thoroughly discredited, and we've embarrassed them twice already."

"Twice?" Cate said.

"First, when the CDC declared it a non-event after the EPA dragged them in, then with the chimp testing."

"But this isn't about us, it's about the evidence."

"They won't listen, Cate. They won't look."

"But...even to you? They have to. You're—"

"Nobody. They've dropped me as well."

"I don't believe it!"

"I was invited, in no uncertain terms, to resign while I had the chance."

"Oh, no."

"Which suggests pressure's being applied. They wouldn't have dumped me just for bringing them another report. Bill told me

I'd made the wrong people seriously pissed. Even he's been getting hints to distance himself from me, and fast. I'm afraid I'm very much 'persona non grata' now."

"I'm so sorry."

"I tried to call the CDC directly. They haven't returned my calls. What a surprise. I'm waiting for the other shoe to drop."

There was silence for a while as they digested George's news.

"There's nothing else for it then," Marc said. "We have to go public. Call in the press. The whole three-ring circus…maybe leak the chimp results first? Let it build a bit? Let them come to us, then drop the bombshell."

Ayden's phone warbled urgently. It was in meeting mode so someone in his contact list was claiming an emergency to get through.

[Query.]

[Voice call from Tess James.]

[Accept.]

"Have you seen the news?"

[No. What's up, Tess?]

"They're saying you were paid off by Bond University for bringing them in on the Hunt River project. Here's the link." Ayden's phone confirmed receipt of a link packet from Tess.

[Thanks, Tess. You know it's a set-up, right?]

"I know it. It's that damned Genenco again."

[I've got to check this out, Tess. I'll call you back.]

"Okay."

[Thanks, Tess. Bye.]

Ayden looked around the room. "I think we all need to see this."

He flicked the link to each of them.

The newsreader was full of certitude, swelled with self-righteous indignation. Apparently, their investigative reporting team, following up on an anonymous tip about a pay-off from Bond University, had uncovered a Swiss bank account in Ayden's name with $150,000 dollars in it.

"I thought Swiss accounts were numbered," Cate said.

"They are," Marc said. "But obviously they aren't going to let the facts get in the way of a good story."

"That's the other shoe then. Well, they certainly won't listen to us now," George said. "Actually, it helps to explain my reception at the EPA and the CDC."

"You know I've been set up?" Ayden said, looking at George.

"I know. You care less about money than anyone I know. But it doesn't matter. The mud'll stick."

"More to the point," Marc said, "even the damned media won't give our story any cred now. Bastards have done a thorough job on us. Perfect timing. We must be bugged."

"I've had a scanner cycling through the full spectrum, looking for transmissions ever since we moved back," Ayden said. He looked out the window, could easily see at least twenty windows in the distance. Exhaling sharply, he pulled the blind.

● ● ●

"Dammit," the sniper cam operator in the bare, fourth-floor apartment three hundred yards away said.

● ● ●

"And I thought I was being sufficiently paranoid," Ayden said.

"Don't sweat it," George said. "There are people out there who eat, sleep, and breathe covert surveillance, and they don't all work for the government."

Silence. The knowledge that everything they said and did may be recorded certainly stifled conversation.

"So what do we do now?" Marc said.

"Is it safe to even discuss it?" Cate said. She looked at her husband. "God, I don't know if I can go through this again. First Gardner, now Genenco."

Ayden felt hollow. "I'm sorry, Mom. I'll make it right." Somehow.

She looked at him through tears. "It's okay, I just…" She ran from the room. Marc tried to force a smile and went after her.

Ayden and George looked at each other.

"Who's Gardner?" George asked.

"Long story. I thought we were being careful but, one thing we can be confident of now, there's no such thing as enough paranoia. One moment."

Ayden flicked an icon in the corner of his specs. [Command. Access house speakers. Play music, jazz genre, random selection, set volume to forty.]

Ayden sat huddled in the middle of the room with George, their heads only a few inches apart. They spoke quietly, only just able to hear each other over the music.

• • •

"Me too," the sound operator said to his colleague who was packing up his sniper cam.

He'd placed a microphone in the ceiling of the Walker house, just above a bookshelf. Anyone in the living room would have had to climb a ladder to get a book from the top shelf in order to notice it. He'd laboriously run a hair-thin fiber optic cable through the wall, behind a downpipe, across several back yards and up into the roof of the house four doors down, wherein the transmitter was located.

It had taken him three consecutive nights to install.

• • •

"George," Ayden said. "After all your years in the government, is there someone you could call? Someone you could talk to? Informally?"

George Reyes regarded Ayden silently.

Ayden continued. "Someone for whom covert surveillance, deception, and elaborate set-ups are all in a day's work?"

"Yes," George said finally, directly into Ayden's ear. "There's someone I could talk to but listen up. If they take an interest, they'll run the show and they'll run it their way. What happens next will be out of my hands." George drew back a fraction to fix Ayden with his eyes "And yours, Ayden. These guys aren't tolerant of independence."

Ayden gazed back at George. "Do we have a choice?"

George let out a deep sigh. "I guess not."

Friday, March 10th

Ayden and George met in Sausalito for lunch.

They discussed the replacement team taking over the bee project. Testing had been judged a success, and they were well on the way to full-scale implementation in time for spring.

They discussed the latest nano-tech news headline. A team in South Korea had managed to grow carbon nano-tubes up to a meter long. Finally, there was a materials advance that might just be strong and light enough to make space elevators possible, 140 years after the concept was first proposed. They fantasized about the rejuvenation of space exploration that would be possible if getting into space cost three dollars per kilo rather than fifty-thousand.

They discussed George's plans for his retirement.

They discussed the weather.

As they said their good-byes and shook hands, George reached up to pat Ayden on the shoulder. As his hand descended, he slipped a small piece of paper into Ayden's top pocket.

On the train ride home, Ayden glanced at the piece of paper.

"*Tomorrow. 2:00 p.m. Grand Hyatt Hotel. Lobby. Frank Howarth.*"

Saturday, March 11th

Small clusters of armchairs were scattered around the lobby of the Grand Hyatt Hotel. Ayden chose one in a prominent position, sat down and waited. The chairs weren't as comfortable as they looked, but it didn't matter. He was there for less than five minutes when a tall, white-haired man in a nondescript business suit approached him.

"Hi, Ayden. Frank Howarth. Pleased to make your acquaintance. I've got a room upstairs where we can talk." His lined face carried an easy smile.

When they got to the room, Howarth didn't reach for the handle. He knocked instead. A young woman in a severe, navy blue suit opened the door.

"Please come in," she said, standing aside. Closing the door behind them, she turned to Ayden. Her face was utterly serious, devoid of warmth or emotion.

"Please shut down your phone and give it to me."

"You want me to take it all off?"

"Sorry."

She put his cuffs, neckband, earpiece and specs into a black suitcase sitting on the bed and left the room.

They sat at a small coffee table near the window. The curtains were closed. Frank cracked open a bottle of sparkling water, poured two glasses and took a long sip.

"So, what can I do for you, young man?"

"You're George's contact?"

"Never met the man."

"But you're government intelligence?"

"Nope." He gave Ayden a wide smile. "Frank Howarth, 43 Pennyworth Lane, Washington D.C. Look me up. Political consultant, lobbyist, purely private sector. Great job. Access to all sorts of people in government and business without raising eyebrows. I talk to people. I float around the edges of things, particularly where the government and corporations come together.

If something's going on that I think certain people should know, I let them know."

"But really you report to one of the government intelligence agencies? You are the person I should be talking to?"

"Hmm…see now, here's how it works. I don't know you, and in my business it doesn't generally pay to show too much of your hand, especially to people you've just met. So why am I here talking to you? Because I trust the guy who asked me to talk to you. Clearly, he trusts your guy."

"Okay."

"So you need to ask yourself one question. Do you trust your guy to broker this conversation? Cause you won't get any assurances from me. Wouldn't be worth anything if you did. Decide if you trust your man and don't waste too much time 'cause I'm supposed to be attending a cocktail party across the other side of town in a couple of hours."

"I trust George Reyes."

"Good. Good. One more thing, I'm not a scientist, but I'm told you know your stuff so it shouldn't be a problem. Specialists can look over your information later on if need be. Just tell me, in plain terms, what you think's going on."

Ayden did. He told Frank all about the bees and the mosquitoes and the evidence they had unearthed connecting the mosquito virus to human infertility. Howarth let Ayden finish without interruption, then fired question after question at him. For hours. He missed his cocktail party. Finally, the questions wound down.

"Okay, so what it boils down to is a lab accident, only the folks at Genenco don't want to get blamed and maybe they're happy to see the virus spread anyway."

"It looks that way."

"Or the virus was deliberately cooked up and spread."

"I have no evidence either way."

"What's your gut feel? Off the record."

"I'd like to think it wasn't deliberate."

"Hmm. So if it's decided this is worth looking into, even though the CDC has given the all clear, assuming your data pans out and we think your conclusions are plausible, what sort of action would you expect us to take?"

Ayden was nonplussed. He truly hadn't expected that question.

"Er, I'd expect the government to use its resources to obtain additional data. I would expect urgent action to contain and counter the virus internationally. I'd imagine a press conference, mobilization of the world's medical and scientific communities."

"And what about Genenco? Prosecution?"

"I doubt you'd find enough evidence, but at least you could keep a close watch on them."

"All right, so here's what I'm going to do. I'm going to get some people to look through your data. You have it with you?"

"Yes." Ayden handed Frank a memory card.

"Thanks. Now, if we think this all looks the way you say it does, you can be sure the people who need to know, will know."

There was a lengthy pause. When it was clear that Howarth had said all he was going to say, Ayden prompted him.

"Then what?"

"Then, my young friend, what happens next will be out of your hands. You understand that, right?"

"Yes, I do."

"Cause I've been asked to be completely up front with you, to treat you nice, but you need to understand that if the people I talk to decide to step in, then they'll be running the show."

"I understand."

"Do you? What I read about you indicates you have a bit of a habit of going off and doing things your own way. The guys who run things might decide to follow a different 'what next' than the one you would choose."

"I understand."

Frank Howarth leaned closer, fixing Ayden with his eyes.

"So here's what I need from you now. I need you to clam up. No more contacting the authorities. No more asking questions. No more collecting data."

"I'm here because there's not much more we can do without backing."

"I know that, Ayden, but your actions following this conversation will be watched closely, and they'll indicate whether or not you can be trusted to play ball."

"I understand."

"You keep saying that, but it's not quite the same thing as saying 'I agree' or 'Okay, I'll do as I'm told.' Is it?"

Ayden paused, weighing his words carefully. "I'll give the government every opportunity and every assistance."

"Hmmm, I guess that'll have to do for the moment. What about your parents? Can you speak for them?"

"Yes."

"Definitely?"

"Yes."

"And what about Tess James? She's been shooting her mouth off all over the place, hasn't she? She doesn't seem to be the type to sit quietly in the corner."

Ayden hadn't mentioned Tess by name, had tried to keep her out of it.

"I'll talk to her."

Howarth just held his gaze steadily.

"I think I can convince her to be patient."

"You think you can. Can you convince her or not?"

"I can…for a while at least."

"Good 'cause the people I work with aren't fond of loose cannons. They're good people, Ayden. They're on the side of the angels, but the world's a terrible messy place so these guys tend to be very hard-nosed. Now you and your family's credibility is shot to hell already, the last thing you need is the U.S. government on your back as well."

"I believe you have made your point, Mr. Howarth."

"Good. Okay, Ayden, that's it for now. I'll be in contact."

Howarth rose, opened the suitcase and handed Ayden his phone gear. He offered Ayden his hand.

They shook. "How do I contact you, if necessary?" Ayden asked.

Howarth smiled with genuine mirth. "Oh, I'm easy to find. Frank Howarth, 43 Pennyworth Lane, Washington, D.C."

Friday, March 17th

As Cate looked around the breakfast table, a tiny smile tugged at the edge of her lips. The kitchen was warm and smelled of fried eggs. Her chickens had started laying. The room itself was simply furnished with bamboo board cabinets and white countertops, but as usual, it was filled with lush greenery. Marc, lost in his specs, sipped coffee. She'd years ago given up trying to talk to him until he'd finished at least his first cup. Ayden and Tess were quietly chatting about some guy with a weird avatar. It was a warm, domestic scene and Cate drank it in, used it as a foil to ward off her worries.

"So Dizzy's stumped?" Ayden said to Tess.

"He won't admit it but that's my guess. He knows every exploit in the book for obscuring identities but this is seriously old school. He's quite offended. Seems it's just not chill, for a hacker to hide behind legal confidentiality," Tess said.

"Doesn't sound like Jaal's your average info hacker then," Ayden said.

Marc finished his coffee. Specs clearing, he looked around as if noticing the others for the first time. "Water anyone?" he said, heading for the fridge.

The doorbell rang. Marc frowned. "I'll get it."

● ● ●

Ayden watched his mother's face tighten at the sound of the doorbell. Damn. Until that moment, she'd actually looked relaxed for the first time in weeks.

Marc stood in the kitchen doorway. "There's a young woman at the door asking for you, Ayden."

It was the same serious-faced young woman Ayden had met with Howarth at the hotel. She even wore an identical suit.

"You remember me," she said.

"Yes," Ayden said. He could hear quiet footstep behind him. He glanced around to see his parents looking worried and behind them, Tess wearing a quizzical expression.

"Please call me Lucy. Frank apologizes for the lack of communication, but we believe it's best to proceed with caution."

"Well, come in, Lucy."

"Actually, I'm here to ask you to come with me. All of you. Expect to be away all day."

"Where are we going?"

"We can discuss it further in the car."

"I'll need to get my bag," Cate said. "Marc, would you lock up, please?"

A few minutes later they were getting into a large, tan-colored four-wheel drive with very dark window tinting. It was a seven-seater so Ayden and Tess climbed into the back row of seats, leaving the middle row for Cate and Marc. The young woman sat beside the driver, immediately turning to face them as the car pulled from the curb.

"We couldn't talk in your house," she said. "As you are aware, you have been under surveillance on behalf of persons unknown. We figured they'd know about us anyway, so we cut their access to you as best we could, but they had some pretty sophisticated equipment so we couldn't be sure we'd got everything."

"Where are you taking us, Lucy?" Cate said.

"To a secure research facility. It'll require a short flight."

"Where?"

"We'll be there in about two hours."

"Yes, but where's this facility?"

"Sorry, ma'am. I can't say."

"I think we have a right to know where we're being taken."

"Sorry, ma'am. I'm not permitted to tell you any more about the facility except to say that Frank will be there to answer further questions."

Cate stared at the young woman, who calmly returned her gaze. Eventually, she gave up and pointedly looked out the window.

They drove to a small airfield where the vehicle drove into a hanger. Inside was a small jet aircraft.

Cate stopped at the stairs to the aircraft and turned to Marc.

"I'm not so sure we should be stepping into that plane. No one knows where we are," she said.

"If you prefer not to join us, ma'am, I'm authorized to have you returned home," Lucy said.

"No problem, Mom," Ayden said. "You and Dad go home. Take Tess with you. I'll be fine."

"No way," Marc said. 'Cate, take Tess home. Ayden and I will go—"

"Hey, sorry but I'm going," Tess said. "I want to know what the hell's going on."

Lucy stepped forward. "Ma'am, I can assure you that—"

"You can't assure me of anything!" Cate turned on Lucy with a ferocity Ayden had never seen in her before. "Or you won't. You've repeatedly refused to answer our questions, and I see no reason why we'd believe you anyway. Now please let me have a conversation with my family."

Lucy raised a hand in surrender and took a step backwards.

"Mom, if they intend us harm, or even just to lock us away, I'm sure they could do it whether we voluntarily step into that plane or not." Ayden turned to Lucy. "Is that a fair assessment of our situation, Lucy?"

"We don't intend you any harm," Lucy said.

Cate fixed Lucy with a stare. "I'm not happy with any of this."

"Sorry, ma'am," Lucy said to Cate's back as she led the way up the stairs.

Ayden wasn't surprised to discover that the window shutters wouldn't open. They couldn't see outside the aircraft at all. Hardly a word was spoken throughout the fifty minute flight once Lucy had collected all their phone gear. When the doors opened, they saw they were again in an aircraft hanger. This one was huge, and Ayden noted three fighter jets, two of which were pointed toward the big doors and looked ready to scramble at a moment's notice. There were also several military-style vehicles parked in the hanger.

"So much for the sights," Marc said.

A small door at the back of the hanger opened to what could have been any administrative offices anywhere. The corridor was lined with windows. Venetian blinds blocked most of the view but allowed some natural light into the corridor. Ayden caught vague glimpses of desks, filing cabinets, and people through the blinds.

Lucy led them down a flight of stairs and along another corridor, this one featureless except for a few blank doors. She opened the second door on the right and gestured them inside.

Frank Howarth was sitting in one of six comfortable-looking, black chairs on swivel bases, surrounding a low table. He rose.

"Thank you all for coming. Hi, Ayden," he said, shaking Ayden's hand with a wide smile before turning to the others.

"Mr. Walker, I'm Frank Howarth. Call me Frank. I'm pleased to make your acquaintance. You must be extraordinarily proud of your son." He turned to Cate. "Mrs. Walker, I'm pleased to meet you and I'm really sorry we had to drag you out here but we had to ensure this conversation was completely private."

After a moment, Cate shook Frank's outstretched hand.

"And this must be the irrepressible Tess James. Pleased to meet you, young lady, and yes I do have information for you. Please everyone, make yourselves comfortable. Can I offer you refreshments?" As he spoke, a door opened and a young man wheeled in a trolley. He gave Cate a cappuccino, Marc an espresso, Tess a chamomile tea and Ayden a glass of water with a slice of lemon.

Marc's eyebrows raised. "You're showing off, Frank." It was exactly the order each would typically have made at a café. He sipped his coffee. "Good coffee, too, very good. They're lying to us, Ayden. This can't be a government facility."

"I wanted to meet you at the other end and fly here with you, but I'm afraid I was overruled," Frank said. "I thought a decent cup of coffee or tea might indicate we're making an effort."

"So you're not in charge around here?" Cate said.

"No, ma'am. Not even close. My job today is to give you folks a brief rundown of what we've been doing for the last few weeks, answer your questions as best I can, then introduce you to the person who is in charge.

"First up though, I'm here to thank you on behalf of the United States government. It looks like you've unearthed something our government needed to know about." Frank looked around smiling.

"Uh…thanks, Frank. Can you tell us what they plan to do about it?" Marc said.

"Some, yes. Basically, over the last week, they've confirmed the veracity of your claims. They had shifts tag-teaming twenty-four/seven. They've re-analyzed the CDC data and collected more from other sources. It all agrees with your work…" Frank paused.

"Yes?" Cate said.

"…as far as it goes. Problem is, there's still way too much we don't know. Our superiors don't really like making decisions of geopolitical magnitude without all the facts. They need to know more. Is this just an escaped bug or is someone deliberately driving it? Who's involved? Why does it only affect some men? Is the effect permanent? Will it affect the next generation? Can we vaccinate against it? Is it reversible?" He shrugged. "They've got a list."

"All good questions," Ayden prompted.

"And we need your help."

That got their attention.

"See, we've got scientists—good ones—but time is clearly of the essence. Getting our scientists up to speed with the stuff you've been working on for months, without access to you…well, that takes more time than we have. And in some cases, the decision makers are gonna ask for best estimates—guesses. Now, I'm told good guesswork needs a good scientist with a deep understanding of the subject…the sort of feel which only comes with time and our guys reckon your work is first class. We've also checked your credentials and, prior to the current controversies you were highly regarded. So we got the security boys to take a look at you and they gave you a provisional bill of health. Actually, as a bonus, by using you folk we bring more resources to the project without widening the circle of those in the know."

"But surely you're going to go public with this anyway?" Marc said. "So aren't security clearances irrelevant?"

"I suppose we will, but there are lots of different ways of going public. Picture a couple of headlines. Here's one: 'Scientists advise vaccination against flu virus linked to fertility issues in some men.' Sound okay? Calm? Sensible? How 'bout this: 'U.S. plot to sterilize developing nations.' That sound like one to avoid?"

"Yeah, okay," Marc said. "But that's just plain silly anyway."

"Is it? Affected men aren't totally sterilized, are they?"

"No."

"So anyone who can afford IVF can have babies."

"Most, probably, yes."

"So middle class Americans are okay, but poor Muslims in Africa are…what?"

"Hmmm."

"And the organization you accuse of being behind this mess is a pillar of the U.S. biotech industry. A company with numerous links to the government."

"Okay, I get your point," Marc said.

"So, we're gonna want to control just how we release the information coming out of this project."

"The fastest way to derive a solution is to open up the problem to the world's science community," Cate said. "Science works best, by far, as a collaborative venture."

"But, once again, there are more and less sensitive ways of going about it. Besides, I think you'll be impressed by the capacities of your country's bio-defense resources."

"So, let's go," Ayden said. "Let's meet your boss and get started. We have some initial recommendations—"

"Whoa, that's great, but first you've got to decide whether you want to sign-up for the project," Frank said.

"That's why we're here."

"Excellent. Well, then I've gotta get a lawyer in here to explain the provisions of the Official Secrets Act and the agreement you all need to sign."

"You're kidding," Cate said.

"Sorry, but this is the government. You've got to expect some bureaucracy."

• • •

The lawyers' brief overview took more than an hour. Essentially, what the Secrets Act amounted to was that the government could declare any body of information not widely known as critical to national security. That done, it became treason to divulge it to anyone without authorization. All the agreement really did, according to the lawyer, was provide evidence that they had been fully informed about the Act and that the project was covered by it.

Nevertheless, they demanded time to read the documents.

"We'd be mad to sign this," Tess said after a quick scan.

"If that guy's telling the truth, then it doesn't really make any difference," Cate said.

Ayden asked for the lawyer to be brought back in.

"Who do you work for?" he asked. "The military?"

"Actually, I'm employed by the Department of Justice. Why do you ask?"

"You told us that the only effect of this document is to provide evidence that we've been informed of the Act and the declaration. We'd like that advice in writing please, in your capacity as an officer of the court and on DOJ letterhead, and we'll want to lodge a copy with our lawyer before we proceed."

The lawyer hesitated. "There may be one other effect," he said. "The agreement notes bio-terrorism as one of the threats we face. That introduces the possibility that the government could use its powers under the Homeland Security Act of 2002 as amended…"

"Ahhh, lockup without trial," Marc said quietly. "Charming."

"If they did that you could, theoretically, be held without trial for an extended period, yes," the lawyer said.

"For how long?" Tess said.

"Well…since the 16th amendment, ahh…there isn't a particular limit, but there are all sorts of checks and balances…"

"Indefinitely," Marc said. "This just keeps getting better. Ask Frank to come back in here, would you please?"

• • •

"They didn't tell me about the terrorism clause," Frank said on his return. "If they had, I'd have advised them to discuss it up front. I'd have told them that you folks are too smart for legal 'sleight of hand' games."

"It doesn't exactly get the relationship off to a good start, does it?" Cate said.

"No, ma'am, it doesn't, and I'm sorry about that, but I'll tell you two things I know for a fact. First, the laws really do apply whether or not you acknowledge them by signing the paperwork and second, you don't go any further unless you do sign. This isn't the little league, folks. You decide. Get in, be involved, give yourselves a place at the table, or go home and do something else. No one will hold it against you."

Ayden tried to persuade the others to walk away, to leave it to him to pursue. Cate and Marc made it clear they wouldn't, under any circumstances, leave Ayden to go it alone.

"This is the most important thing happening on the planet," Tess said. "I don't want to sign that stupid document, but if the alternative is not knowing what's going on then…what the hell, I guess."

In the end, they all signed.

"I know it's not an ideal situation," Frank said. "But that's just the way the world is. The good news is, now that you've signed, I can introduce you to Major Henri Soulandt, who runs the show here. I don't wanna steal the major's thunder, but there's some serious science being done here and good progress has already been made."

Major Soulandt entered the room from a featureless door at the rear. They had expected a military uniform blazoned with insignia. Instead, she wore civilian trousers and a blouse. She was, perhaps, in her late thirties or early forties, looked fit, and carried an air of quiet determination. She wore no makeup, and her hair was tied back in a neat bun.

After introductions all around, each accompanied by good eye contact and a firm handshake, the major led them to the back of the room, toward the door she had entered by.

"I'm going to take you straight to where the action is. I don't need to tell you that none of you would be seeing any of this if it weren't considered critical. I have a responsibility to deliver information within the tightest possible time frames so I'm going to give you a very quick guided tour then assign you all tasks. I'll be speaking with each of you individually over the next forty-eight hours or so but in the meantime, I need to get you straight to work. Questions?"

"The next forty-eight hours?" Cate said.

"Yes. I'm going to ask you to live on campus for a few days, at least."

The major opened the door to reveal a small, empty room with an elevator door in the far wall. They entered the room and she closed the door behind them.

"I can assure you of comfort and privacy. We employ a group of very unmilitary scientists here and some of us have lived on campus for years."

"Oh, I can understand that but…"

"Write a list. I can send someone to your house to pick things up if you like or we can purchase what you need."

As they filed into the elevator, the major fixed Cate with her eyes. It was as if she had instinctively decided that, if she could appease Cate, then the rest would fall into line.

"I'll do my best to make you all as comfortable as possible, Cate, but you know how important it is that we don't waste a moment, don't you?"

The lift descended for a long time.

Cate sighed. "I'd prefer it if your people didn't enter our home."

The elevator doors finally opened to a long, featureless corridor.

"You guys just love corridors, don't you?" Tess said.

The major didn't respond. No doubt they served some security purpose. As they neared the far end, it was opened by a young woman in another conservative, but unmilitary, outfit. She gestured for them to enter a large office.

"This is Mother," said the major with a brief smile. They looked at "Mother" who must have been all of twenty years old. She had bright orange hair, which seemed determined to escape a bun high on the back of her head.

"It's Mother who really runs things down here. Keeps everyone sane, so we can keep working."

"Hi, y'all," said Mother in a soft, lazy southern accent. "What the boss means is I'm a gofer. You need anything, you let me know. Anything about this place gets your goat, you let me know. I get to spend the good taxpayer's dollars making the cave livable." She smiled, her cheeks dimpling charmingly.

"This is the boss's office," Mother said, pointing to a door to her left, "but she only goes in there when the brass comes down for a visit or to make calls. This," she paused to open the door on the opposite side of her office, "is the cave."

The door opened onto an astonishingly large space. It was much larger than the aircraft hanger somewhere above. They stood, literally speechless, gazing into the "cave." Their hosts, evidently used to first-time visitor's reactions, simply waited. What caught Ayden's attention most was the light. Though they were clearly some distance underground, the cave was bathed in bright, natural sunlight. He peered at the ceiling far above, at the big, bright panels.

"Fiber optics," the major said, standing next to Ayden. "One hundred percent natural sunlight. For those of us who don't go upstairs for weeks on end. Makes all the difference."

And the plants. There was greenery everywhere. Huge planter boxes contained masses of shrubs and ferns, even small trees. Clearly, there had to be another entrance to the complex, an entrance capable of admitting trucks. It was more than bearable, it was really quite pleasant.

"Mother will show you around, get you settled in," the major said. "I'll see you in forty minutes, over lunch, to give you your assignments."

Greenery divided the huge space. Near the door was a series of areas defined by overflowing planter boxes and small collections of outdoor furniture. The overall effect was of a large garden with nooks for people to escape to, and read or talk. The ambience was reduced by the concrete floor, but even in this, some effort had been made. There was a dull olive tint in the concrete that softened the grey.

Mother had played tour guide before. With obvious pride, she pointed out features of the space as they walked through. To their left, behind a wall of green, was a gymnasium area full of exercise equipment. Beside it was a glass-encased squash court. Next, a series of enclosed meeting rooms, mainly glass walled but offering "one hundred percent audio insulation" according to Mother "and visual privacy, if needed." The glass turned opaque white at her gesture.

They reached the far end of the huge space. Outside an enclosed kitchen, cafe-style tables and chairs offered a relaxed dining environment. The back wall was essentially a series of doors. "Those go into the various labs," Mother said. "Then down the right-hand side, those doors lead into the residential rooms. I'll show you to your rooms. We can sleep over 150 people down here, and everyone gets their own room. I'm told we could sleep two to a room if we had to, but it wouldn't be as much fun as it sounds."

The rooms were tiny. There was room for a bed on one side and a small table and chair on the other. Mother had found two adjoining rooms for Cate and Marc. "Sorry, we're not geared for families down here," Mother said, then turning to Marc impishly, "but you wouldn't be the first feller to sneak down the corridor in the middle of the night." She cast a tiny flicker of a glance toward Ayden as she said it.

Tess rolled her eyes.

Ayden and Tess were not far away, four or five doors apart. Bathroom facilities were shared.

Throughout the "tour" Ayden noticed how many people cast curious glances their way. "You're what passes for celebrity down here," Mother said. "Probably half our people have been re-tasked to investigate the Pandora virus and they know y'all discovered it."

"The Pandora virus? There's an optimistic name," Tess said. "I'd have thought we were hoping we can get this virus back in its box."

Mother looked at Tess blankly.

Lunch consisted of platters of cut sandwiches and fruit. They met the major in one of the glass fronted meeting rooms, which could have been in any corporate office anywhere. A dozen steel and faux leather chairs surrounded a large, honey-colored conference table. "First up, phones. I'm afraid you don't get your own back 'til you leave. Down here, you'll use these." She gestured to the phone gear Mother was handing to each of them.

"You've loaded them with our personal configs?"

"That's not going to happen, I'm afraid."

"You've gotta be kidding," Marc said, around a mouthful of sandwich. "You have any idea how long it took me to train my gear?"

"Sorry, but the military system's completely incompatible."

"That's smart. Make everything more expensive and more difficult to develop apps for."

"And stay secure. From machine language through operating system to all apps, security built in from the ground up. But it won't be as bad as you think."

"You reckon your gear's that much better?"

She smiled, shaking her head. "Look, your neckband has to interpret the electrical signals in your throat nerves, in conjunction with the tiny vibrations it picks up when you subvocalize even completely silently, right?"

Marc nodded.

"Same thing for your cuffs. The inertial sensors tell them how your wrists are moving and the electromagnetic sensors tell them what you're doing with each individual finger. All very tricky 'cause we're all different, but the clever bit isn't the hardware, it's the software interpreting the inputs, like those dogs they train to pick up on the subtle body language clues to warn their epileptic masters they're about to have a fit."

"You're saying you have smarter software."

"Not as such, but what would happen if your gear didn't have to rely on its own processing power to train itself? What if you accessed ten or twenty grand's worth of supercomputer time while you're at it?"

"Per person?"

"Easily. Maybe more. One thing we've got down here is petaflops to burn. You won't believe how quickly your new gear will learn you."

Marc sat back. "This'll be interesting."

"That's the attitude we're after. Now, your new phones, of course, operate on an entirely stand-alone network. You'll have limited read-only access to the outside grid but it'll be heavily filtered. That means delays. So, if your phone suddenly seems slow, be patient. It means your request is being filtered, then routed all over to obfuscate its source. Even your info requests could provide eavesdroppers with hints about your work, so one query from this facility will appear to come from a Los Angeles internet café, the next will come from a University in Prague or a private residence in Osaka."

"Sounds slow."

"Can be."

"Just like the bad old days…the World Wide Wait." Marc smiled and examined the sandwich platter again.

"It gets worse," the major said. "Your request may get bounced. That's our filters refusing to pass your query on to the grid. The bounce will be logged but if you think it's important and shouldn't have been blocked, you'll need to flag it for one of the security team to evaluate."

"What about calls? Voice messages?" Cate said. Her plate was empty. She hadn't touched any food.

"Sorry, were not allowing voice for the moment."

"We can't even let people know we're okay?"

"Not with voice. We're in full lock down for a while. You'll be able to send and receive text but be warned, all outgoing text will be read by a member of the security team and some may get bounced back at you."

"That's ridiculous. You can't cut us off like that," Tess said.

"We're trying not to." The major sighed. "We're trying to compromise, to give you an acceptable level of access. Believe me, it would be a lot easier to just cut you off completely. If you really feel the need to talk to someone then we'll arrange for you to do it from my office but I'll be there listening and I'll have my finger on the cancel icon."

"I want to tell George Reyes where we are, or at least in whose care we find ourselves," Cate said. "I don't trust you so much that I'm happy being down here without anyone knowing about it."

"That's reasonable. You can't give him details, of course, but you can tell him you're spending a few days with associates of the friend he introduced Ayden to. That do?"

"I guess it'll have to."

"Good. Now, to your assignments...Cate, a vaccine is in the works."

"That's quick."

"Thank you. We've gotten pretty good at rush vaccines since sixteen."

"'Sixteen?" Tess said.

"Christ's Hammer," Marc said. "The Superbowl Killers."

"Ah," Tess said.

"I can remember it," Ayden said, peeling a mandarin. "I'd just started school and was loving it, but then I had to stay home and wasn't allowed outside to play."

"But weren't you still living in Australia then? I thought Australia managed to stay uninfected."

"Sort of, except for flight seven-seven-one. All countries closed their borders, but the planes already in the air had to land somewhere. Flight seven-seven-one was quarantined as it landed and the Royal Australian Army Medical Corps set up a field hospital on the runway and contained it."

"How many died? Of the people in the plane?"

"Don't know," Ayden said. "Dad?"

"There were 232 people on the plane," Marc said. "One hundred and six of them died. Believe it or not, that was a record outcome. In the States, the average was over 50 percent."

"And sixteen of the eighteen cultists themselves," the major said. "Of the ones who actually sprayed the Ebola virus onto the popcorn and sold it to the crowd."

"Less diligent medical attention, presumably?" Tess said.

"I can imagine they weren't the highest priority," the major said. "But really it was just because they got infected days earlier when they were setting it all up. Apparently, they were told they'd been given a vaccine and the initial symptoms they were experiencing were just reactions to the vaccine, but it was a lie... fortunately for everyone else."

"What do you mean?" Tess said, her food forgotten.

"Patient zero turned up at Emory Health on the Monday, nauseous, dizzy, severe headache, and joint pain, the symptoms of any number of infections. Luckily, they had enough beds to keep him for observation and when they saw the skin rash and red eyes, the doctor immediately quarantined him together with herself and her medical staff. She enforced strict barrier nursing and contacted the CDC.

"The hospital became an official controlled zone within two hours, and the CDC supplied a vaccine before the end of the day. Although they were struggling to discover how the patient had contracted the disease, the medical community was starting to congratulate itself on a textbook-perfect response to its appearance when a second, obviously very sick man presented himself to the press gathering outside the hospital cordon.

"Live on television he explained what they had done. Thousands of the Godless, who chose to watch violence on Sunday rather than going to church, would die. They had only hours to repent before they'd be judged by their maker."

"And the blind panic that followed kick-started the quarantine?" Tess said.

"Well, yes, most Americans went home and locked themselves in, but the real reason it was fortunate was that Ebola incubates for at least a week and then it takes another week for the rash to show up."

The major looked at each of them in turn. "As it was, we lost over twenty thousand people. Can you imagine how bad it would have been if the virus had been circulating widely for weeks before patient zero turned up?"

They were silent for a while.

"Of course!" Marc said to the major. "You know everything about it. It's part of your job."

"Sure. It's why places like this exist."

"So you can tell us. Why didn't the CDC's vaccine work? Were the rumors true? Was it a weaponized virus? One of ours?"

The major hesitated. "Not as such."

"What's that mean?"

"Look, some of the details are still classified."

"We've got security clearance, don't we? And we signed your damned document, and it's ancient history anyway. As a sign of good faith?"

The major turned to Cate. "Persistent isn't he?" She sat back. "Okay, I give in, but this conversation is covered. It goes no further." She waited until she had a nod from each of them.

"It wasn't a U.S. bio-weapon, but it was deliberately bred to be vaccine-resistant by a U.S. bio-defense worker. The cult obtained the virus from Africa, where it hadn't quite finished exterminating the last of the wild chimpanzees and gorillas, and Reynold Atkins, a senior technician at the Arizona E & N University lab, was able to get hold of a sample of the vaccine. They set up a private lab for him and he got to work."

"The University employed a religious extremist?" Tess said. "Clever."

"He wasn't when he was screened, but his marriage failed and the divorce got ugly and there was a fight about the kids, etcetera, etcetera."

The major shook her head in disgust.

"Atkins hit the bottle," she said. "He spent a year as a functioning alcoholic. Then he nearly sent a vial of the vicious Spanish flu virus that wiped out fifty million people in nineteen-eighteen to a Level Two lab. E & N covered it up. They'd already filed two incident reports that year. They kept him on, but he had to stop drinking entirely and submit to random alcohol breath tests. After a year of inebriation, the abrupt detoxification was torture. He joined AA, read web sites, and books, tried psychotherapy.

"He turned to religion. Finally, he found the comfort he was looking for. Apparently, he'd always been a bit obsessive and within a year, he was a member of Opus Dei.

"It wasn't until he met the woman who would become his second wife that he'd even heard of Christ's Hammer, a small offshoot of Opus Dei, determined to reform it, to correct its indulgent ways. I've seen photos of her. Tiny and very pretty. She, of course, refused to adulterate herself or marry outside her church. He joined without hesitation."

"So places like this were set up to control bio-defense research a little more closely," Cate said.

"Oh, yes. It had gotten a little out of hand following the anthrax scare in oh-one. Half a decade later there were over twenty thousand people working in more than four hundred labs across the country."

"So how many places like this are there?"

"Now that, I really can't tell you. Back at the turn of the century, planes and bombs were the weapon-of-choice for people wanting to attack us and most threats were neutralized without Joe Public ever finding out. These days, it's the grid and bio-weapons. You have no idea how many seriously scary situations we shut down before they do any real damage."

She paused and her eyes met each of theirs in turn before she continued. "We're good at what we do. We already have a candidate vaccine for your Pandora virus and we're about to start testing."

"Impressive," Cate said. "Reverse vaccinology?"

"TCR peptide in IFA," the major said.

"In eight days! How?"

"Hold it, hold it," Marc said. "You gals think you can mix a little English in there for the dummies?"

"Sorry, love. I assumed they'd scanned the virus's genome and reverse engineered its surface proteins to find a good target for the vaccine. It's really fast but doesn't always work. Turns out, they're using a T-cell receptor peptide in an incomplete Freund's adjuvant instead."

"That's English? Sorry I asked. Now I'm hoping I don't need to understand. So let me guess, you're going to test this peptide vaccine on soldiers?"

"Correct. They don't seem to mind the idea of avoiding un-planned pregnancies and we committed to keeping frozen semen for them and funding IVF if it all goes pear shaped. I don't think they gave much thought to how unpleasant IVF would be for their partners. In any case, the peptide vaccine is on track but by the time we roll it out worldwide, there are going to be a lot of people already infected. We need to develop a therapy. We need to trick the patient's immune system into recognizing his sperm as domestic again or find some other way to stop it producing sperm antibodies. Think you can help, Cate?"

"Yes, of course, I can try."

"Good. One thing though. Here, we don't 'try,' we 'do.' Next, Marc. So far, you've modeled the spread of the virus. We need to extend that analysis significantly. What does it mean for birth-rates, all around the world? What does it mean for world popula-tion and demographics in fifteen, twenty-five, and fifty years? We need to model that for varying assumptions about the success and timing of a worldwide vaccination program…and with different assumptions about IVF uptake and reversibility of the virus."

"Sounds pretty straightforward. That all?"

"No. Then we need you to work with the econometrics guys to make predictions about the impacts on the world's economies and major industry groups. I've got some heavyweight statisticians on loan but they say your modeling treats some of the non-linear complexity in ways they're struggling to confirm. They have used brute force computational power to prove out your basic spread predictions but they say the extra layers of complexity from vac-cination, reversibility and sociological and economic reactions would benefit from some of your finesse."

"Nice of them. By yesterday, right?"

"The day before would be better. The good news is you will have access to a ten exaflop super-computer to assist you."

Marc's eyes lit up. "Excellent. Where do I go?" He stood up.

"Steady feller. Ten more minutes, then you can go play with your new toy."

Marc sat down.

And fidgeted.

"Tess, I want you to help Marc. He's used to Cate interpreting the science for him and defining the source data he needs, but she'll be very, very busy."

"Yeah, okay." But Tess sounded uncertain. It sounded to Ayden is if she was wondering what she had got herself into.

Finally, the major turned to Ayden. "I've saved the doozy for you. It's impossible, of course, but from what I read, it's right up your alley. One of the fun things about working for the government is they always get around to asking the Rumsfeld question."

"The...sorry, I don't follow you."

"Donald Rumsfeld. Turn of the century bureaucrat. Secretary of State the third time we invaded Iraq. No, sorry, the second time."

"Doesn't ring a bell."

"Lucky you. He's famous for saying 'there are things we know we know, things we think we know, things we know we don't know... and things we don't know we don't know'... or something like that. Anyway, our military and political masters, who have to make decisions, lose most sleep about the last item in that mangled list. The things we don't know we don't know. They're the things we can't take into consideration, the things we can't plan for. Often the things that turn around and bite us hardest on the butt."

"Okay."

"Your job, Ayden, is to tell us what we don't know we don't know. About the virus," the major paused, "and about Genenco and William Hanford."

"Okay."

"Initially that means intensive debriefing. Days of it possibly. The good news is that then we get to brainstorm. Much more fun. Then we research and prune our wild ideas."

"We?"

"Pandora is my only priority. And aside from generally keeping people on track, the Rumsfeld question is top of my list."

Monday, March 20th

"How long do you suppose we're going to be stuck down here, Ayden?"

They were sitting in his room. Tess sat cross-legged on the bed. Ayden sat on the room's only chair, tipping it back on two legs, the chair back leaning against the desk. His feet were propped up on the bed.

"I have to say," Ayden said, "I'm amazed how pleasant they've managed to make this place. I mean given that it's run by the army and it's hundreds of feet underground."

"I guess they had to make some effort or the scientists would walk," Tess said.

"Maybe, but they seem to like the sense of purpose."

"That and not having to beg for the next research grant. It's still a hole, and we're still cut off from the world. You may be happy living a hermitic existence, but I have a social life—had one anyway. They bounce half my messages, and the other half are so delayed you can't have a proper conversation."

Ayden laughed. "I was told they'd had to add an extra person to the screening team just to keep up with your texts."

"Serves them right. You can't tell me you're enjoying yourself?"

"Not the debriefing. That was a day and a half of mind-numbing tedium but the brainstorming is actually fun. Henri wheels in one team of specialists after another and we wring them dry. She chairs ruthlessly. Ideas are tabled, noted and excluded from further discussion, then we kick 'em out, assign priorities for analysis and drag the next lot in."

"Henri, huh."

"C'mon, Tess, she's just doing her job. How's working with my dad?"

Tess brightened. "He's pretty amazing actually. He claims he's just a number cruncher, but he keeps digging into the biology 'till he really understands the problem, then he does magic. The other number jocks scratch their heads and complain about degrees of freedom and matrix drift then your dad makes these mental leaps. By the time the others have satisfied themselves his approach is valid, he's applied it and moved on."

"The man's an artist. Anything interesting coming out of it?"

"Some scary stuff. Like, without vaccination or IVF, world population could drop too low to sustain a modern industrialized society within ten generations."

"But IVF does work, and there is a vaccine in the works."

"Even so, it'll take ages to roll it out. If we can't reverse it for those already infected, world population could drop a third by twenty-one hundred."

"Doesn't sound so bad to me."

"Maybe, but the economists are seriously green around the gills. How's your mom doing?"

"No progress, so far, reversing it for those already affected, so we better hope none of our Rumsfeld what-ifs come true."

"Ah, yes, the wild speculation you and the major are indulging in, while the rest of us work. Pray tell, great prognosticator, what ills could befall us?"

"Ouch. Well, here're a few samples for you. What if, instead of facing a natural dispersion of infected insects, we're dealing with an aggressive, coordinated effort to spread the virus as far and fast as possible?"

"That doesn't actually sound so far-fetched to me."

"What if the virus mutates, becoming immune to the vaccine, or becoming more virulent, affecting 100 percent of people infected like the chimps? Or what if it has a selective effect?"

"Like what, for example?"

"We keep talking about the virus affecting fortyish percent of men at random, but, of course, it isn't really random. Some men's immune systems, in fighting the virus, overreact and recognize their own sperm as foreign. Others don't. So what if it's not only deliberate but also targeted? What if someone's trying to take a particular type of person out of the gene pool?"

Thursday, March 23rd

By unspoken agreement, they gathered each day around noon in the garden dining area. The chairs and tables were cheap, white, fiber-plastic but they were comfortable and encircled by big planter boxes full of shrubs and ferns. Today, Ayden and Henri arrived first. They chatted as they looked over the selection of wraps, salads and fruit.

"I can't find any direct evidence of selective effect," Ayden said. "And it's not through lack of data." They had access to medical records from hospitals and IVF clinics worldwide.

"But?" Henri said.

"At first glance, it seems members of the armed forces may show a greater impact from the virus compared with the general population."

That got her attention. "Have you controlled for other variables?"

"We're working on it."

"How would you explain it?"

"I can't. Yet."

"Could it be a red herring?" Henri said. "An artifact of different record keeping between the military and civilian medical establishments?"

"Very possibly," Ayden said. "We'll certainly try to determine if that is the explanation."

Henri looked at him, a hint of a smile tugging the corner of her mouth, eyebrows raised.

"Sorry, Major, make that *we will determine*."

Her smile fell away. "Do you have any particular reason to be concerned?"

"You've told me to consider the possibility that someone is deliberately spreading the virus. That raises the question of whether it was a lab accident at all. And, if it was engineered, then it may have been targeted at a specific group."

"Not at all easy to achieve," she said. They found a table and sat.

"How do you know? Have you tried?" Ayden gave her a very direct look. "What do you really do down here when you're not analyzing mosquito viruses, Henri?"

They hadn't noticed Tess approach. Henri was just opening her mouth to reply when Tess dragged a chair out from under the table and sat with her sandwich. Ayden looked up. Marc was at the salad bar.

"Would it be too much to ask why you haven't gone public yet?" Tess said. "You tell us you're on the verge of being able to produce your vaccine in huge volumes, using both government and private-sector facilities, but there's not a whisper of it on the grid. Why isn't there a massive publicity campaign and queues of men lining up to buy insect repellent?"

"Hi, Tess. Nice to see you, too. Believe it or not, I don't make policy. I just send information upstairs to the people who do."

By now, Marc was sitting down to join them. "It's a fair question," he said. "The government does seem to be dragging its heels."

"Do you really believe they would? Ayden, what do you think?"

Ayden regarded her for a few moments. "You're using the malaria outbreak, aren't you?"

"He does it again. We just can't slide one past you, can we?"

There was no restriction on news feeds into the cave, and they had all noticed the news. With global warming and ubiquitous travel had come a very worrying spread of malaria and the mosquitoes that carried it. Recent reports revealed outbreaks in North America. Now, the U.S. government was leading a worldwide push to eradicate the virus from the world as had successfully been achieved with leprosy, polio, and smallpox.

"Hang on a sec," Marc said. "We were reading about malarial outbreaks in New Mexico the day before we came down here."

"Yes, but a week after your son spoke with Frank."

"You were preparing the ground. Now I'm impressed. That shows great forethought for—"

"For the public sector? How nice to be appreciated. It was my suggestion, actually."

"It's not enough," Tess said.

"How so, Tess? We've contracted every available vaccine manufacturer on the planet to help us adapt our production method for large scale output. By the time we finalize testing, they'll be ready to go. In a couple of months, we expect they'll be pumping out ten million doses a day. We've mobilized the world's medical infrastructure to deliver maximum possible coverage worldwide. We've even given unvaccinated people reason to be wary of damn mosquitoes. Horse-crap humility aside, I'm damn proud of what we're doing. What more do you want?" It was the first time Ayden had seen Henri loose her icy cool composure.

Tess shook her head. "Anyone would think you people were allergic to the truth. It'll take too long. Ten million doses a day? That's pissing in the ocean. There are eight billion people on the planet. It'll take years at that rate. Way, way too long."

While Tess spoke, the major had recovered her own composure completely. "It won't take anywhere near that long. Will it, Marc?"

"No. We only need to vaccinate the men, actually men of child-bearing age or below. That cuts it down to around two billion. So, if we can actually vaccinate at a rate of ten million a day…six months, give or take."

"Thank you, Marc. And you're right to question the logistics of delivery and actual vaccination. That'll be the bottleneck, rather than production volumes. Thankfully, we no longer need to use needles. Aerosol injectors can be used a million times, and it takes a completely untrained person all of 5 seconds to give someone a shot."

"I don't understand why you can't just be truthful. You'd get a more urgent response, better and quicker coverage. Besides, your cover story means you have to vaccinate women and old men as well."

"Give us some credit, Tess. We're saying there are different doses for women and old people."

"And they get, what, saline solution?"

"With this logistical cost and effort? Actually it was decided to have a go at really eradicating malaria. That's the beauty of it. The cover story is real. The threat is real, though we've exaggerated the urgency. The women get real malaria vaccine, the men get both. The latest RTS recombinant malaria vaccine is showing 88 percent efficacy, which is high enough to do the job if we get enough coverage. It may be hard to believe, but we really are trying to do the right thing."

"You still haven't explained why you can't just tell the truth. Is honesty completely irrelevant to you? Don't people have a right to the truth?"

"They also have a right to personal safety and a civil society if we're able to deliver that as well. And a job. And regular meals." She glanced over the table which was crowded with plates of food.

"What are you taking about?" Tess said.

"Ayden?" Henri said.

"Panic in boardrooms and on the street," Ayden said.

"Go on," Henri said.

"Riots around hospitals and vaccination vans. Theft of shots. Looting. And the financial markets will go berserk. Investors will do math like Dad's, well not as accurate or elegant as yours, of course, Dad." Ayden smiled fleetingly. "But the virus has been active for months. They'll know that population levels will take a dive even with the vaccine, and that consumption will follow. They'll dump stocks. It'll be a major share-market rout followed by a real and deep business cycle downturn. Businesses will avoid capital investments, expansion plans will be shelved. Unemployment will get even worse."

"Two thousand and nine. Nineteen twenty-nine," Marc said. "It could last decades."

"Recessions don't last decades." Tess said in a quiet voice.

"Sure they do. In Japan, the Nikkei reached thirty-nine thousand back in nineteen eighty-nine. It spent two decades below half that level and didn't get back to its old highs until twenty-seven. Over a third of a century. Believe me, Tess, this thing's gonna get messy, and that's assuming the vaccine works perfectly. If it doesn't work and the effect isn't reversible, we'll be looking at complete economic collapse."

"Of course, the truth will emerge eventually, if only as revealed by the demographics themselves," Ayden said.

"True, but later, more gradually," Henri said. "This way we give ourselves breathing space. If the extent of the truth emerges slowly, markets and businesses get time to adjust. People get time to adjust. And..." she looked around "it gives us time to come up with answers. Time to work out how to reverse it. Time for IVF businesses to ramp up."

There was silence for minutes. Finally, Tess locked Ayden in her gaze. "Do you think we should lie to the world?"

"It's not my call to make."

"If it was?"

"I don't doubt people would panic and do a lot of harm. I don't like hiding the truth, but...sometimes it's...necessary."

For Ayden, whose parents had hidden the truth from him for most of his life—who had been hiding the truth about himself from Tess for months—the conversation was getting uncomfortable.

He was about to continue but suddenly Tess's expression went from skeptical to concerned and she was looking past his shoulder. Ayden turned his head to find his mother standing there, pale faced.

"It doesn't work," she said in a near whisper. "The vaccine doesn't work."

• • •

That night, near midnight, Ayden knocked quietly on the door to Henri's room.

"Who is it?" The question came immediately. Henri hadn't been sleeping.

"Ayden."

"One moment." At least two minutes later, the door opened. She was wearing a silk robe and slippers. Her hair fell softly around her face. It was the first time he'd seen her without a somewhat severe bun. She may even have been wearing...well, perhaps not lipstick but some sort of lip gloss or moisturizer.

"This is an unexpected pleasure," she said. Her smile was warm, ambiguous. "Come in."

She stepped aside, allowing Ayden to squeeze past. He found himself in a room three or four times the size of his own, and without a bed. Instead, a large desk dominated the room, surrounded by a very comfortable looking chair on one side and two much less comfortable looking chairs on the other. There was another door at the back of the room.

"My official office is too disconnected from everyone else," she explained. "Its amazing how many informal but valuable conversations I've had since I let people know about this room. Not usually as late as this though."

"I'm sorry, I can come back—"

"No. Come through," she said opening the other door and leading him into a second room of similar proportions but very different décor. This was furnished as a comfortable sitting room, two armchairs opposite a couch. Old-fashioned hardware paintings on the walls, seaside landscapes. A rug on the floor. Even flowers. There was yet another door, ajar. Ayden caught a glimpse of a bed before he sat down.

"This is nice."

"Ah, yes, the trappings of power. I even have an en suite bathroom. Would you like the full tour?"

"I'm sure you spend enough time down here to warrant some comfort and privacy."

"Hmmm." She regarded him silently for a few moments. "So what can I do for you, Ayden?" she said.

"I…didn't want to ask at lunchtime, but how long do you intend to keep us down here?"

"I appreciate your discretion. Today was dramatic enough as it was."

He just looked at her.

"Clearly, we've got a lot more to do and this is the best place to do it. Is it really so bad down here?"

"No. Not really, but the longer I stay down here, the more apparent it is that you don't really need us, and I feel…disconnected down here. With the news that the vaccine doesn't work, it feels like I'd be more effective up there. Finding out what's really going on."

"There's still the secrecy aspect to consider, Ayden. Things are at a very delicate stage right now." She stopped. Ayden was smiling and shaking his head.

"Doesn't wash, Henri. There must be hundreds of people topside who know the truth or have a pretty good idea."

"People who can be trusted to preserve the administration's secrets, at least for a while."

"The world ignored us when we were shouting it from the rooftops. Why worry about what we'd say now."

"Now you know about this facility. About our plans." A hand waved vaguely.

"You didn't have to reveal this facility or your plans to us. We're assisting but I don't think we're making a fundamental difference. As a reason for illegally imprisoning U.S. citizens, it's looking a bit thin." Ayden was entirely calm, but it was time to get to the bottom of it.

"You're not prisoners here. You agreed to come down."

"Okay. So can we leave, please? Tomorrow morning?"

She hesitated. "We're going to have to ask you to stay with us as our guests for just a little longer."

"It's a pleasant enough cage, Henri, but we're still here whether we like it or not, and largely incommunicado."

"Hardly incommunicado, Ayden. You have full access to information feeds, and can text out anytime."

"If I'd wanted the official line, Henri, I wouldn't have disturbed you at midnight."

This time there was a longer pause. "Okay. I should know better with you. You're stuck here for the duration, but no more so than I am."

"Why?"

She sat forward. "Is it so unpleasant to be here?" She smiled warmly.

"Why, Henri?"

There was a long silence as they regarded each other. Henri smiled a wry smile, sighed and sat back. "Why do you think, Ayden?"

"You know." It wasn't a question.

"We do. Well, I do. No one else down here does."

"How?"

"We found your data chip."

It had everything on it. Almost. Gardner's experiments. His parents' unwitting role. Ayden's genetic modifications. It didn't, however, include Cate's more recent research on Ayden's telomeres.

"You searched the house after I met with Frank."

"We, in the broadest sense of the word, did. Yes."

"We never noticed, we never even suspected."

"Heads would have rolled if you had."

"I thought you were paying too much attention to me personally."

"Don't sell yourself short, Ayden. You are a fascinating, attractive person. And possibly the smartest person I've ever met."

Ayden's mind raced. The disk was all about him. It didn't mention Hanford. "Would I be as fascinating if you didn't know that I had been genetically engineered?" he asked, stalling for time.

"Yes, Ayden, you would."

There was no reason not to tell them about Hanford now he knew his secret was out. Every reason to warn them what they were up against.

"There's more," he said.

"We thought there would be."

"What else do you know?"

"I've been briefed about the contents of the data disk. As far as I know, that's all anyone knows," she said.

"You need to know about Hanford."

"He's GM, too?"

Ayden only winced a little. "Yes."

"Hmmm, that fits. He's as smart as you?"

"You were testing me."

"We were. Is he as smart as you, Ayden?"

"At least."

"Damn it. We really needed to know that. How many others? Who else?"

"According to Hanford, lots of attempts. Lots of failures—psychoses, autism...lots of neutral results. No other obvious successes. What *fits*?"

"You know you were under intensive, sophisticated surveillance, right?"

Ayden nodded.

"We traced it back to Genenco, we think. So the question is...this virus, is it a deliberate thing? Is Genenco behind it? Is Hanford?"

"Did you trace the surveillance to Hanford, or just to Genenco?"

She held up a hand. "We can't prove anything yet. Genenco's firewalls are extraordinary. But the surveillance firms, the ones watching you, have history with Genenco and we have uncorroborated testimony linking them to Hanford's office. So?"

Ayden sighed.

"Hanford and I have been playing cat and mouse for months now, but I don't know," he said. "Yet."

"C'mon Ayden, your guess is better than anyone else's."

"It's very possible."

"I'm going to take that as a 'yes' for the moment. Is he taking advantage of a lab error or did he set out to engineer this?"

They were both leaning forward, tense, talking very quietly. Ayden was suddenly aware that Henri gave off a fresh soapy fragrance.

"It's very possible. We need to find out."

"Agreed. That's your focus from now on."

"From in here? It'll take too long. Can't you just…grab him?"

"No, Ayden, we can't just 'grab him.' He's way too prominent, way too powerful."

"Unlike us."

Henri sighed. "Yes, Ayden. Unlike you."

"Then let me out. I'll find out."

She leaned back in her chair. "Believe it or not, if it were up to me, I'd let you out. I'd trust you to do the right thing then come back to your family. But it isn't up to me. Your best shot is down here digging through the data."

"I don't think I can do it from down here, Henri. He's too clever, too well prepared. I need a free rein."

"That's not going to happen, Ayden, and you know it. But short of opening the door for you and committing career suicide just to watch you get captured ten feet away, I'll do what I can to help."

"Well, I can't say I'm surprised," Ayden said.

"Now we're coming clean, you know we'll need more debriefing sessions, don't you? About Hanford, your contact with him. Every conversation."

"Oh yeah. Can't wait."

"Fun, fun, fun."

"Hmmm."

She leaned forward again. "So, now you know I'm not try-ing anything sneaky—and this is the last time I set myself up for rejection—would you like to stay the night, Ayden?" Her hand reached out. "Here with me?"

Ayden closed his eyes and sighed. He opened his eyes and looked into Henri's and gave her hand a gentle squeeze. "No, Henri," he said softly.

"Too old and haggard?" She took her hand back.

"Hardly. You're beautiful."

"I wouldn't whore myself for my government, you know."

"I know you wouldn't."

"Tess?"

"She would be hurt."

"And you want to keep your relationship with your jailer clear?"

"I guess…something like that, too."

She looked down into her lap. "The people I work for have no idea the sacrifices I make."

"I should go now."

"Okay, Ayden. Debriefing tomorrow."

• • •

She didn't look up as Ayden let himself out, as she heard the click of the door.

"Henrietta, you're a goddamn moron," she muttered to herself. "And you've been down in this hole for too long." She didn't move for several minutes.

Then she sighed, reactivated her phone, and started to dictate her report. Her only solace was that, as she would mark it super urgent, her immediate superiors would be woken up to read it. They wouldn't get any sleep either.

Friday, March 24th

Ayden headed for his room, his mind whirling. Some unknown number of intelligence, military and government personnel knew his secret. The circle of knowledge would ever widen. Ayden's life would never be the same again.

Again.

He stopped walking. Now there was no reason not to tell Tess. He was afraid of how she'd react but she had to hear it from him. He'd go to her first thing in the morning. He looked around. He was standing outside her door.

He tapped softly, then a little harder.

"Who the hell is it? What time is it?"

"It's Ayden."

"Ayden? Just a sec." He couldn't help thinking there was something funny about the amount of time he was spending knocking on doors this evening. This morning, really.

Tess's eyes were wide, but she'd definitely been asleep. She had, however, taken a moment to brush her hair. "Is everything okay?"

"Everything's fine. Can I come in for a minute?"

"Yes, yes." She drew him inside and sat down. "What is it, Ayden? Do you have any idea what time it is?" she said looking around for her specs.

Ayden grabbed the chair, moving it close to the bed. He sat close, facing her and, very quietly, almost whispering, he told her everything. Almost. He didn't mention his potential longevity.

● ● ●

Ayden woke, as he always did, at 6:00 a.m. He'd never needed an alarm clock, even after only a few hours sleep. He lay unmoving. Telling Tess had been a release.

He was still wearing yesterday's clothes.

"Just where do you think you're going?" Tess murmured as he tried to slip off the bed without disturbing her. He felt her arm tighten across his chest. When he had finished his story she had lain back on the bed and said "Can I have a hug please?" They'd fallen asleep that way.

"I need a shower. Sorry, I tried not to wake you."

"We finally sleep together, and nothing happens! Then the best you can do is, *I need a shower*. You aren't going anywhere, Buster." Tess rolled on top of him and kissed him deeply.

This would make things even more complicated. He should stop her. His response surged through him. His clothes were flung off. Her night-dress disappeared.

The feeling of her bare skin pressing against his was electric. He stopped thinking.

· · ·

It might have been an hour later. It might have been several. "Now I really need a shower. And so do you."

"Not yet." She inhaled deeply. "Besides, we smell good. Strong, but good."

"We'll be late."

"What are they going to do, fire us?"

He laughed. "Good point."

"I'm not finished with you yet."

Major Soulandt's debriefing could wait.

Monday, March 27th

The cross-examination sessions ran from 7:00 a.m. to 11:00 a.m. then 4:00 p.m. to 8:00 p.m. Between 11:00 a.m. and 4:00 p.m. Ayden was expected to continue his research.

He was sitting with Marc in the corner of the room Marc shared with three other analysts. It was quite spacious and there were desks for another two people but they'd managed to fill the place with clutter.

It was even worse when Ayden slipped into their shared virtuality. Marc's old-fashioned chalkboard competed for attention with virtual whiteboards and 3D graphs. How people whose role it was to analyze information could work in such a disorganized workspace, Ayden couldn't fathom. His father, however, was clearly enjoying himself.

"The answer is 68 percent," Marc said. "That's after controlling for every extraneous variable we can think of."

"Across all the different defense forces?"

"Yep, Army, Navy, Marines, Air Force, Homeland Security, even the Intelligence services—those I could get data for anyway. They vary a little from service to service but between 63 and 71 percent of our defense personnel are made infertile by this virus. That's against 41 percent in the broader population."

"Are you sure?" Ayden said. "How's that possible?"

Marc grinned. "Beats me. I get to pitch the curve balls here. I don't have to try to bat them too. Apparently there's a team somewhere confirming the source data, but the analysis is rock solid."

Ayden shook his head. "Well, it's going to be interesting investigating this one. Do we have data for foreign military personnel?"

"Some. Same results so far." Marc's grin faded. "They still interrogating you?"

"Yeah, but it isn't particularly aggressive or unpleasant. It's just tedious. In the mornings, I get 'Larry' for about an hour then 'Tony.' Larry focuses entirely on Dr. Gardner but I'm sure he knows more about Gardner than I do so it's pointless. Tony wants to know everything I know about Genenco and William Hanford, which also isn't very much. The afternoons are more interesting though. He calls himself "Carl" and he isn't interrogating me at all. He's running me through psychometric tests."

"Hmmm," Marc said. "No prizes for guessing–" Marc stopped as Ayden held up a hand.

[Voice message from "Mother",] his earpiece said. Ayden still hadn't gotten used to the way the military issue phones overrode all personal filters with command messages. The only way he could miss the message would be to physically remove his earpiece and specs. ["Hi Ayden, Major Soulandt would like to meet with you straight away in her office."]

"Sorry, Dad," Ayden said. "Henri wants me in her office."

"She say what about?"

"No."

"Her office, not her room?"

"Yes." It was the first time any of them had been called to her formal office.

"Okay," Marc said. "Uh, send me a quick message when you're done."

• • •

Ayden sat in the outer office chatting to Mother.

"She won't be long," Mother said. She smiled and tucked a wisp of orange hair behind an ear. She seemed not to notice when it escaped again almost immediately. "Did I tell you? We may be getting fake grass for the cave! You know, the stuff the football stadiums use? I've been tucking away some funds from the budget for months and apparently the Navy stadium in Annapolis is going to replace theirs so we may be able to grab it."

Ayden wasn't really listening. The door to Henri's office was slightly ajar so he could hear her voice from inside as it became raised.

"What do you mean only a dog? If a dog can get into the vent shafts, then other things could too. It'd take kids from Wilson all of 45 minutes to get here by dirt bike. I want a detailed plan on my desk by the end of tomorrow, including costing and staffing. I want both shafts examined tomorrow as well. Entrances, crawl spaces, plant room, the lot. I want to know exactly how the dog got in and how we're going to stop any more unexpected visitors. I want twenty-four-hour electronic surveillance of the entrances and shafts within a week. I can't believe there isn't any at the moment. And I want a schedule for manual inspections to be carried out at least weekly. I also—"

Suddenly the door closed, shutting off all conversation from within. Mother continued to enthuse about her latest cave beautification project for several more minutes, then finally told Ayden he could go through.

Henri looked tired. "Hi, Ayden. Sorry to keep you. You wouldn't believe some of the crap I have to deal with." She sighed. "You didn't hold back, did you?"

Ayden raised an eyebrow.

"I'm not talking about information. I mean on the tests."

"Actually, they were almost enjoyable after the tedium of the cross-examinations."

"Well, you've really got yourself noticed now."

Ayden just looked at her and waited.

"You're completely off the charts. They ditched the normal aptitude tests after the first session. Switched to Mensa Society questions. You blew them out of the water."

Ayden smiled.

"The last set of questions weren't even part of the normal Mensa tests, they were Society brain-teasers. This is, of course, secret but you are officially the most intelligent person ever tested."

"They haven't tested Hanford."

"No, though I suspect he'd tactically underscore himself anyway, stay under the radar. Why didn't you?"

"So you'd take Hanford seriously. So, do I get a certificate? A Mensa application form?"

"What's really got their attention though is your SI score, your social intelligence. A great many Mensans find it difficult to hold down a job and they certainly aren't over represented in political or corporate leadership. Extreme IQ scores are usually associated with markedly low measures of social or emotional intelligence. The people who rise to the top have high, but not extreme, intelligence combined with excellent control over their emotions and awareness of others' internal states."

"And?"

"You maxed the SI scores as well."

"I see. Well, that was Dr. Gardner's objective."

"It seems he succeeded. Now my superiors are really scared of both you and William Hanford."

"That's understandable."

Henri looked distinctly uncomfortable. "And they're getting suspicious about your inability to work out why the U.S. military is being singled out. Maybe you're stalling."

"I suspect that it's as simple as a particular phenotype being overrepresented in the defense forces, but why are they concerned anyway? The virus has no impact on their ability to do their jobs, just their ability to have children."

"Over 60 percent of serving soldiers come from families with a father or grandfather who served, Ayden. This could be a long term but directed attack."

"I didn't know that but, even though the data isn't comprehensive, it's not just in the U.S. that the military is overrepresented. It appears to be a worldwide phenomenon."

"That's supposed to make them feel better?"

"Maybe it's not such a bad thing."

"Don't say that again, to me or anyone. I'm serious." Henri stood and began pacing around the small office. "They really want us to work out what's going on."

"You know I'm doing my best. I want to know as much as anyone else does."

"I believe you." The emphasis was on the "I." "I've been told to do whatever it takes to find the answer."

"Waterboarding won't help, Henri."

"Luckily for you." She smiled fleetingly and lapsed into silence. Ayden waited. The silence stretched. Finally, she spoke again. "One more thing…" There was another long pause and she sat down again.

"Oh hell, sometimes I hate this job." She looked angry and embarrassed. "After what happened between us…or rather what didn't happen…well, anyway." She paused again and took a deep breath, leaning forward and locking eyes with Ayden. "After I propositioned you, this is the last request I want to make, but I have to. Look, the thought hadn't even crossed my mind."

"I believe you, Henri. The answer is no."

"I haven't–"

"I won't provide a sperm sample."

Henri collapsed back into her chair, sighing. "Of course, you'd guess." She closed her eyes. "I'm sorry but you have to."

"No."

"If they really want to, they can just take one. I'm told they can extract samples directly from your testicles under local anesthetic."

Ayden tried not to wince. "You told them I'd say no."

"I did."

"Well, do what you have to do then, but I'll not willingly provide my germ line cells for you to experiment with and reproduce at will. I won't even have a child naturally without careful consideration."

"I'll tell them you'll think it over."

"If you like."

"I'm buying you time, Ayden, that's all. And not much of it."

"Thanks, Henri."

"Use it."

As Ayden returned to the lab, he replayed the meeting in his mind. There was something odd about it. Something Ayden couldn't put his finger on.

Tuesday, March 28th

Ayden's eyes opened.

He'd been sleeping deeply, but his mind had been working. He reached for his specs. It was two in the morning.

He had it.

He lay still, pouring over his meeting with Henri, examining his options. Then he slipped out of the little bed he was sharing with Tess and silently pulled on his clothes. He found her lipstick and wrote a short note on the make-up mirror lying on the corner of her desk. He slipped out the door, leaving his military issue phone gear on Tess's charge pad.

Ayden made his way through the corridors to the cave. At the threshold, he paused, listening intently.

At first the silence was absolute. Then he heard it, quiet footsteps coming his way. Ayden shrank back into the shadow of the door, flattening himself against the wall of the corridor. Motionless, he watched as the night guard walked past the doorway. Ayden crept forward, listening as the footsteps moved away. He risked a glance around the doorway. The man was just disappearing around one of the huge planter boxes. Ayden slipped silently in the other direction, pausing briefly when he reached the cafeteria area then ducking down the corridor leading to the labs.

If he was detected, he would simply feign sleeplessness and proceed to his workstation. Outside the door to the lab he usually worked in, Ayden paused again to listen. When he was convinced that there was no one else moving around in the corridor, he continued past the lab door. Reaching a blank door at the far end of the corridor, Ayden pressed his ear to the door to listen. Unable to hear any movement within, he opened the door and stepped inside, pulling the door silently closed behind him.

From this point on, Ayden would have great difficulty explaining his actions if caught. He had no legitimate business in the plant room.

The room was pitch black and Ayden had no idea of its internal layout. He would have to risk the light. He shrugged out of his shirt and lay it along the crack at the bottom of the door, hoping to minimize light spilling into the corridor beyond. He felt around for the manual switch and flipped it.

As Mother had proudly explained during their initial tour of the facility, the entire subterranean base could, if necessary, cut off all connection with the outside world. They could even survive without external ventilation for up to a month, if necessary. The plant room held racks of massive oxygen bottles and banks of carbon dioxide scrubbers. In normal circumstances, however, the base was ventilated from the surface. Running fans was tremendously cheaper than continuously replacing the activated carbon that the scrubbers required and topping up oxygen bottles. Ventilation also resulted in a more pleasant smell.

Nearest the door was a simple workstation where technicians could interrogate the plant systems if the network was locked down. There had to be maintenance access to the ventilation shafts somewhere…there! It looked a little like the sort of hatch one might expect to find in a submarine. The door was small, up off the floor with rounded corners. In its center was a wheel. He spun it open. As the door opened a crack the machine noise from beyond became shockingly loud. Ayden quickly pushed the door shut again and spun the wheel to seal it.

He went back to the workstation and began opening drawers. Ayden smiled. How like the military to be fully prepared. In the bottom drawer was not only a flashlight, but also spare batteries. Examining the flashlight, Ayden saw that it even carried a spare bulb. He switched the flashlight on and the room lights off, then retrieved his shirt.

Back at the maintenance hatch, he spun the wheel again, stepped through and closed and sealed it behind him as quickly as possible. Turning down the narrow corridor beyond, Ayden paused to peer through narrow slit windows into the workings of the ventilation system. He could see the huge fan blades spinning past the window. Alongside each window was another access hatch half the height of the one through which he'd entered the corridor. Continuing, he came to the last window and access hatch, past the last of the spinning fans. Inside, the noise was deafening and the wind was substantial. To his right was the huge fan. To his left were three large, circular grids in the wall, each with a solid hatch bolted fully open. Examining them closely, he found what he was looking for.

The middle grid showed traces of blood and fur. The dog he heard Henri discussing had evidently fallen down this ventilation shaft and come to a halt against the grid. This had to be the shaft which permitted external access to a medium-sized animal, hopefully large enough for Ayden too. As a bonus, the screws had recently been loosened, presumably to remove the carcass. It was relatively easy for him to undo them with the butter knife he had swiped from the cafeteria.

Ayden clambered into the shaft.

There was no way he could re-secure the grid from inside. Even upside down, Ayden could not have reached the screws from inside with the grid in place. Clearly the base had been designed to keep people out rather than in. Once his searchers checked this room there would be no doubt which way Ayden had gone. He would have to be well clear by then.

He peered up the shaft. As far as his flashlight would show him, the shaft continued in a straight line at about 60 degrees. Steep, but thankfully, not vertical. Ayden surmised that it was yet another form of redundancy. Angle each shaft to a different location on the surface.

He slipped out of the shaft and removed his shoes, socks, shirt and trousers. He tied his shoelaces together and slung them over his neck, also tying his clothes around his neck. He wedged the flashlight securely into one of his shoes.

Ayden climbed back into the shaft and used the natural stickiness of his skin against the stainless steel shaft to edge his way upwards. Lying on his back, he pushed against the opposite side of the shaft with his hands and knees and shins to ascend. Fortunately, the wind in the shaft kept him relatively cool. Nevertheless, he had to rest regularly, or the sweat of his exertions would make his skin too slippery.

After an age climbing, he looked down, carefully angling the flashlight. The shaft wasn't vertical and the grill at the bottom was open. Nevertheless, a fall from this height could maim or kill him. He pointed the flashlight up the shaft. No sign of the top.

Shortly after his fifth rest, Ayden's head bumped against something. He looked up. It was a simple cross-grid. He'd reached the top of the shaft. Certainly, the quartered gaps were big enough for a dog, but would there be room for him? Very carefully, Ayden removed the shoes and clothing from round his neck and hung them over the opposite side of the shaft. Pushing hard against the opposite wall of the shaft with his knees, he tested the cross-grid, but it held firm, and he couldn't see any bolts or screws.

Ayden's head fit through the gap easily but his shoulders and chest would be a challenge. Once, as a child, Ayden had seen a contortionist fit his entire body through the unstrung head of a tennis racquet. Ayden was no contortionist but thankfully he didn't have the bulk of the military guards. They would never have fit. One arm first, then his head. He trailed the other arm behind to maximize the angle of his shoulders. It was a very tight fit. Pushing desperately with his knees and shins pasted against the shaft, Ayden breathed out, minimizing his chest diameter, and inched upward. He took another breath, locking himself in place, then exhaled again to squeeze up another inch.

Finally, he managed to extract his leading shoulder and in a moment he was out.

Ayden sank to the ground to rest, grabbing the flashlight to look around. The lip of the shaft from which he emerged was about four feet off the ground and was surrounded by a circular concrete wall. Hinged fully open, a heavy motorized hatch was clearly able to seal the shaft entrance. Ayden angled the flashlight upwards. The diameter of the room was small, but the walls were about fifteen feet high. Under the overhanging roof, Ayden could see heavy grillwork covering openings which obviously allowed the air in. He sighed. There was no way to get up to the openings. Nor would he have any way to get through the grid if he was able to get up there.

But if a dog had gotten in, he should be able to get out. Peering around the base of the wall, Ayden saw the hole and smiled. The construction of the vent opening was rough and ready. They'd basically placed a large concrete cylinder on the ground, then capped it, and for whatever reason a dog had dug under the wall.

Ayden had to enlarge the hole but it could not have been more than fifteen minutes before he was out.

His flashlight off, Ayden stood with his back to the cylinder and observed silently. He had no idea of the time. Without a phone, Ayden was reduced to making guesses that would have been put to shame by his distant ancestors. The landscape was flat, vegetation consisting mainly of low scrubby trees. Ayden was surprised. He had expected to see a lit up base including the aircraft hanger they had taxied into when they arrived. He moved silently around the outside of the cylinder. There it was, but not a light to be seen. It could have been a small, deserted airstrip anywhere, a hanger and a few small buildings in the dark.

He scanned the horizon carefully. Ayden didn't know how big "Wilson" was but if it could be reached by trail bike in less than forty five minutes, it wouldn't be far, perhaps fifteen or twenty miles. Unless the town had a better than average approach to night-time light pollution, he may be able to detect the glow.

There. It was faint but, despite the half moon overhead, the rest of the horizon was unrelentingly dark. Now he scanned the starscape. The glow was to the southwest.

Ayden put his clothes and shoes back on and, painstakingly cautious, started towards what he hoped would be Wilson. He had no idea what security measures were in place on the surface. Almost silent, he nevertheless varied his footfalls in case they had audio or vibration detecting systems tuned to pick up the regular pattern of walking. He stopped every few feet, peering around for any sign of cameras or other detectors.

Finally, he came to a suspiciously consistent set of trees crossing his path. He approached one of the trees and inspected it closely. Ayden found a small, flat disc imbedded in the bark around knee height then another at about chest height. There was another set of discs on the other side. He didn't bother looking but he was confident the next tree would have the same. No doubt, some sort of beam would alert the base if broken. Ayden lay on his back and wriggled under the lower disc.

After another fifteen or so minutes of proceeding carefully, Ayden decided he would have to risk jogging. He settled into a comfortable pace. Fortunately, there was moonlight to guide him and the ground cover was sparse so he made relatively good speed. Nevertheless, by the arc the stars had described, Ayden guessed he'd been jogging and walking for around three hours when the stars were washed away by dawn's light and he came upon the outskirts of Wilson.

Wilson was small and had the look of a town that had once been prosperous but was now struggling to survive. Ayden jogged directly toward the center until he came upon a main road, then followed it in. He found a cash dispenser and deliberated for a moment. There was no point in making an attempt to cover his tracks in Wilson. Where else would he have gone? He passed his wrist under the scanner so it could read the smart chip imbedded beneath the skin, which also read and passed on biometric data confirming it was still within Ayden's body. Ayden withdrew cash to his maximum daily limit, five thousand dollars, silently thanking the people who anachronistically demanded the continued existence of paper money. He then spent a few minutes in the public toilet of the small park in the center of town, cleaning himself up as best he could.

Next, he found an all night diner and convenience store. They sold one-piece phones. Though a very poor cousin to Ayden's phone, they provided basic communication and information access and could be used without other interface devices—as long as you didn't mind audible voice commands or peering and pecking at a tiny touch-screen. They also didn't require any account to operate. Ayden bought a SIM card which would give him 100 hours of grid access.

Wolfing down a large breakfast, and on his third sports drink, Ayden examined the phone's small screen. Wilson was only 43 miles from Denver, which had a huge bus interchange. He'd ditch the SIM card there. He used the phone to order a rental car to be delivered to the diner and finished his breakfast.

Thursday, March 30th

The security guard at the reception desk of the Hampton Building watched the youth approaching his station and felt his face harden with contempt. The kid wore black gear-pants with their multitudinous seal-lock pockets and one of those media shirts displaying a cacophony of colors and a *Zero Point* video clip. The M-shirt threw blocks of color up onto the kid's head, breaking it up into ever changing blocks, an effect accentuated by the patchwork hairstyle and fully mirrored specs.

As he got closer, the guard found he could also hear the discordant soundtrack that accompanied the M-shirt video. M-shirts directed the sound very effectively to the wearer's ears so if he could hear it then the kid must be almost deafened by it. Why couldn't these punks just wear earpieces like everyone else? Saying nothing, he looked at the kid standing in front of him, the tired face and chapped lips. Probably hung over.

"I'm here to see William Hanford," the kid said, loud and insolent.

"Turn that crap off," the guard said softly, knowing the kid wouldn't be able to hear him, that he'd at least have to turn it down if he wanted to continue the conversation. The volume dropped immediately.

"I'm here to see William Hanford," the kid repeated, just as loudly.

"I doubt it, kid. Out. Now."

"I have an appointment."

The guard sighed. At Genenco you followed procedures. "Yeah, yeah, sure you do. Name?"

"Forsythe."

"Forsythe who, you little twit."

"Just Forsythe. Look it up."

The guard consulted his screen then looked back at the kid uncertainly. "Okay, you're here, 'just Forsythe.' You are to go on up immediately, but nobody goes past this station without being recognized. Remove your specs and turn off your shirt display. The face recognition scanner can't filter it properly."

"Sorry. Can't do that. Please phone Mr. Hanford's office for instructions." The youth spoke quietly now, with complete confidence. His demeanor was utterly calm and serious. Something odd was going on but the guard knew better than to question his screen.

He sighed, tapped the screen and mumbled. "Okay, please go up now…sir. It seems you know the way."

"Thank you, I do."

The secretary in William's outer office didn't hesitate. "Please go straight through."

• • •

It had taken Ayden two days to get there by roundabout route, sleeping on busses. Arranging the appointment, though, had been simple. He'd sent a brief text message to the address William set up exclusively for his use.

[San Francisco. Your office. Tuesday. Forsythe.]

As the door to William's outer office closed behind him, Ayden pulled the cheap phone from his pocket. The amber light confirmed he'd lost connection. He powered it down as he walked across the room.

William sat expressionless on one of the low couches, his collarless shirt was black this time, as was his suit. As Ayden approached, William poured water from a jug for each of them. He sipped his, watching Ayden drink thirstily.

"What have you done, Ayden?"

Ayden sat opposite William. "They found out about me, that I'm GM."

"How?"

"They didn't see fit to tell me. Presumably, there have been quiet rumors, despite your father's and your attempts to keep it quiet. Perhaps they found my parents' data."

"What data?"

"They've been analyzing my genome for years, trying to work out exactly what Gardner had done. It was all offline, but maybe the military found a disc. I don't know, but they took a DNA sample from me. If they went looking in the right places, they would have found our unique sequences."

William leaned his angular body toward Ayden. His face took on an intense, fierce expression. "I ask again, what have *you* done?"

"I told them about you, and about what you have done."

William exhaled slowly and sat back. For a moment, he looked sad or resigned, then his face cleared of any emotion. "Why? And why go to them in the first place?"

"Because it was you all along."

"Really?"

"I knew from the beginning that you were only giving me part of the story."

"So you played along, tried to play me. I wondered if that's what you were doing. You were clever though, didn't even explain yourself to your parents or to Tess."

"I couldn't be sure of evading your surveillance."

"Or that they'd be as capable as you at playing the game. What makes you think all this is any more than paranoid fantasy?"

"Patterns. I've set up my virtuality config to help me to see associational patterns. You haven't left any evidence pointing to you, of course, but the only way to complete the patterns is to place you in the middle. You fill the gaps. In any case, you confirmed it when you framed me for the Bond University bribe. It was you who put me onto them in the first place, and you're Jaal, too, aren't you? You even put me onto the poplars. You were planning to set me up from the beginning."

"Contingency planning." A hand waved it away impatiently. "I truly hoped not to have to use it. I'd like to see the way you've set up your virtuality, these patterns."

"Some other time maybe."

"Why did you decide to oppose me? Do you fail to see the need to act?"

"I wasn't opposing you, at least not initially. I was just trying to find out the truth, which you withheld, are still withholding. You admitted you were contemplating letting a virus render nearly half of humanity incapable of reproducing without medical intervention. That's pretty big, by itself. What were you not prepared to admit to?"

"Why are you here?"

"I need to know the rest. It wasn't a lab accident, was it? You engineered the virus from day one. And there's more too, isn't there?"

"Ayden, I recognize that curiosity is the major driver in your life but I don't accept that you've made yourself a fugitive of the government just to satisfy your curiosity. You're trying to achieve something. What is it?"

"I found myself in a position with severely limited options. I've basically been a prisoner. My family still is. I've placed them in the hands of an intelligence community prepared to go outside the law if it considers it necessary."

"So you're attempting to strengthen your position?"

"I have narrated all I know about their activities, in as much detail as possible, which, of course, includes the location of the facility. I have queued the file to be sent to a very wide audience in the government, media and science communities every twenty-four hours unless I roll back the command before each deadline."

"You've bought yourself a little time only. They'll uncover your data caches."

"Yes."

"And you have come to me. The lesser of two evils?"

"You have resources and power. Not so easy to neutralize."

"But you've also made me an adversary."

"I never wanted to but...yes."

"So you are hoping to negotiate some sort of accommodation with me?"

"Yes."

William's face hardened. "Your bargaining position with me also seems very weak," he said.

"I don't think our positions are so far apart. I recognize the problem you're trying to address. I understand why you have done what you have done."

"And you have rejected it, and done your best to undermine my efforts. And exposed me."

"I think that, with enough will, we can accept each other's views and find a path that we can both tolerate."

"Why should I compromise now? I don't see what you have to offer me, Ayden. Why shouldn't I just hand you back to the military and watch you disappear again. You wouldn't get away from them a second time. You know I haven't left any serious evidence pointing to me. I could even use you as a bargaining chip in negotiations with them."

"You tracked us all…all your possible peers. You made contact with me. You revealed yourself to me."

William stared at Ayden. His eyes seemed to be trying to bore into Ayden's skull, to determine what was really going on inside.

Ayden continued. "You consider yourself separate from normal humanity. You were looking for an ally."

"Perhaps. But you chose to oppose me."

"I chose to find out the truth, which you were hiding from me. I haven't really done you any harm you can't recover from. I'm just asking you to tell me what's really going on and allow me the opportunity to try to find a solution we can both work toward together."

"You've done great damage! You alerted them to me. You may have even unraveled a plan I have been working toward for nearly a decade, a plan which might have solved the greatest problem this planet faces."

"I looked at your data. I agree we face a major problem but I think you're overstating the urgency."

"You have no idea," William said, sounded resigned.

"What do you mean?"

William exhaled sharply then stood and walked over to one of the floor-to-ceiling wall screens to stare out at San Francisco.

"Perhaps I should have risked telling you everything." He didn't sound convinced. "I'm already holding back advances in gerontology. Did you really think the field that produced us two decades ago, would have achieved nothing since?"

"I assumed Gardner, and then you, managed to keep it a closely guarded secret."

"We did, but science isn't done in a vacuum. My father didn't discover any of the techniques used to create us. He bought them, in effect, and I've been spending money ever since trying to control the research and buy up the patents but I can't stall progress forever. Medical scientists are on the cusp of breakthroughs, particularly with sirtuin proteins, that will extend the human lifespan significantly. The death rate will drop and population growth will rekindle beyond any estimate you've seen."

Ayden was tempted to challenge William to substantiate his claim but he knew the other man's projections would be rock-solid. He knew he was simply resisting information he didn't want to accept.

"But it will take time for approvals to be obtained and for treatments to be affordable for a wide market," he said. "And in the meantime, we're making progress. We're learning to rein ourselves in. We're finally addressing climate change and we're getting better at preserving marine and wilderness ecosystems."

William turned back to face him.

"No. Homo sapiens are unable to restrain themselves. The carbon dioxide cycle gave them a scare and at the moment they're making an effort. But they won't keep it up. They never do once they perceive the worst of the crisis is past. Forget climate change, humanity has already caused the seventh great global extinction since the emergence of multi-cellular life and it's massively destabilized the nitrogen and phosphorous cycles as well."

Slowly, William walked back to the couch as he spoke, his tall, angular frame silhouetted against the wall screen.

"The result is a dangerously impoverished and fragile ecosystem," William said. "One decent shock and it won't be able to support us in anywhere near the fashion it currently does. And, of course, we may already be supplying the shock. It may already be too late. By the time world temperatures would otherwise stabilize, the Siberian Tundra may be thawing irreversibly, pumping out its massive reservoir of carbon dioxide and methane. Or a major underwater landslide like the Storegga slide 8,000 years ago off the coast of Norway could release the massive amounts of methane currently frozen in sub-sea hydrate deposits. There's about a dozen plausible scenarios that could thrust the Earth's climate past a critical tipping point.

William sat and stared at Ayden. "Do I really need to tell you that methane clathrates contain more carbon than all of the world's fossil fuels combined or that, molecule for molecule, methane is twenty times more powerful at warming the air than carbon dioxide? Or that when an equilibrium is sufficiently disturbed, the offsetting forces holding it in check are destabilized, that wild swings in conditions are experienced before the system restores stability, possibly settling on very different conditions compared with the starting point?"

"I know, I know, the worst case scenario is worldwide crop failure and partial collapse of industrial society. Massive starvation. Much of humanity suffering horribly and dying, the remainder reduced to subsistence living."

"Actually, the absolute worst-case scenario would come from yet another shock – perhaps massive earthquakes and volcanism as the Earth's crust adjusts to changing ice loads, even more clathrate methane released, more warming, then ocean currents failing, turning stagnant and leading to dominance of hydrogen sulphide producing bacteria."

"You're describing the Permian extinction."

"Ninety percent of multi-cellular species disappeared. The microbes very nearly reclaimed the planet they'd dominated for billions of years. Took ten million years for the biosphere to recover."

"Surely that's highly unlikely."

William waved a hand, in a short, sharp gesture. "Oh, true, so let's not worry ourselves about it. Tell you what, I'll go get a gun and a pair of dice. If you throw a double six, I blow a hole in your head, but don't worry it's very unlikely, less than a three percent chance. Do you want to play?"

"Not particularly."

"I'll tell you what is likely. I believe there's around a fifty percent chance that humanity is taking us to the point where, at the very least, many hundreds of millions will die and billions will suffer. I also believe you know all that, Ayden. You know that "making progress" may not be enough."

"But what would even a radical cut in world population do? Fifty years ago, the population was half what it is today and we were doing huge damage even then."

"I agree. Reducing population alone wouldn't be enough by itself."

"Yes, we need to engineer cultural change. Surely, that's something we can work toward together? If we can make people sufficiently aware that their instincts for conspicuous consumption are nothing more than an inappropriate and dangerous form of social – ultimately sexual – competition, they'll rein in some of that consumption. Surely, we can work out how to make conspicuous abstinence trendy."

"Ha! Religions have been trying and failing to sell abstinence for centuries. And I've spent half a decade, tens of millions of dollars of my own money and indirectly hundreds of millions of investors' and the public's money trying to inject some sanity into human culture. If we had enough time, we might just do it too. Actually, I think you'd be better than me at guiding such a campaign. Possibly, we could engineer enough cultural change over a generation or two. But we don't have that long."

Ayden struggled to concentrate, the lack of sleep and stress of the last few days must have caught up with him. He felt spent. "So what have you decided the answer is?"

William fell silent. Finally, he spoke, regarding Ayden steadily. "We have to achieve radical change in one generation. The babies born today must have a very different approach when they are influencing and implementing policy. Humanity has one shot left."

He stopped again, looking at Ayden.

"You did it," Ayden said. "You aren't just cutting back the birth rate. You're doing something else with your virus. What are you doing to humanity, William?"

"For want of better terminology, call it sub-speciation within existing genetic diversity."

Ayden's mind spun. Genetic diversity was a key tool of evolution. When a population became isolated in a new environment, the characteristics of its members often changed faster than could be accounted for by mutation. The new environment conferred advantages to some individuals within the population so that they prospered and their genes became more prevalent in following generations. Hanford had engineered the virus to be deliberately selective. It was engineered to cause infertility in a particular group so that its members didn't pass on their characteristics to the next generation.

He was trying to engineer a speciation event. He was trying to create a new sub-species of human.

"Tell me!"

William's eyes bored into his. "No. I learn from my mistakes."

"You really aren't going to give me a choice, are you?" Ayden's eyes felt gritty, his head ached. "You want me to decide blind."

"There are always choices."

Ayden was silent for several minutes. Hanford watched him steadily, content to wait.

"Okay, I'll join you, but you must agree to tell me everything and to at least listen to my concerns."

There was silence again as William locked eyes with Ayden. "No. I don't believe you. You're immensely intelligent and capable, Ayden, but not a practiced liar. You still plan to try to stop me. You just want to get out of this room first."

Ayden said nothing.

"I want to believe you but I can't. Convince me, Ayden. Please?"

Ayden didn't answer. He knew he wouldn't fool William. He could hardly even think, his head hurt so much. Something was very wrong.

"I didn't think so. Ayden, I'm truly sorry but you've made a fatal error. You have underestimated me for a second time. You've observed that I prefer subtle and indirect methods. I influence people rather than enforce my will. I buy competitors rather than steal their technology. When I come up against dangerous opponents, I undermine rather than assault them. You thought you could walk in here, mine me for information and I would just let you leave to continue to oppose me."

"What are you going to do?"

"I've already done it." A guttural timbre entered his voice. "You confused my determination to avoid thuggish alpha male behavior, my preference for finesse...with weakness." Hanford's voice went hard and flat. "I value life and civilized behavior. I've never had to maim or kill...you assumed that I wouldn't kill under any circumstances...but I will. You are too dangerous an opponent to try to counter subtly."

Ayden's head pounded.

He struggled to speak. "What have you done?" His voice emerged as a croak.

"I've infected you with a meningitis virus one of our businesses developed for DARPA, engineered to be particularly virulent. I have an antidote, in case you were able to convince me you'd changed your mind." He stood, his glance flickering to the huge slab of white glass that was his desk. "Sadly, and I mean that, I won't—"

Ayden moved fast.

He leapt across table.

Too late, he realized that instead of trying to evade him, his adversary simply closed his eyes and clapped his hands over his ears.

The world exploded. Light seared Ayden's eyes. Noise stunned him. He felt his body crash into the other man's but was unable to even reach out for him. He rolled onto the floor in agony. The assault of intense light and the blast of noise triggered pulses of pain that slammed through his skull.

The noise filled Ayden's world, huge and discordant. Tangible. Then, suddenly, it stopped and Ayden's stunned ears could hear only ringing.

The light remained. He kept his eyes squeezed shut against the fierce glare but even through closed eyelids, the light was savagely bright.

The pain consumed him.

Against the steady background pain, intense pulses of pure agony speared through him. Desperately, he tried to think. Normal meningitis caused dangerous swelling of the brain inside the skull. Without medical care it could be fatal. As he and William had been talking, Ayden had been developing a headache. His brain had already been swelling. Meningitis sufferers were acutely sensitive to light and sound. Their medical care included weeks spent in a dark, silent room. The government's Defense Advanced Research Project Agency had long flirted with using light and sound as a weapon. No doubt the increase in blood pressure associated with Ayden's sudden movement when he had tried to jump William, together with the intense blast of light and sound, had pushed his swelling brain across a threshold to an acute level.

If he didn't receive medical attention quickly, he'd suffer brain damage. Left longer, he'd die.

He tried to force the pain into the background. The throbbing made it almost impossible. Keeping his eyes firmly shut, Ayden crawled across the floor. As he did so, the light intensity decreased significantly. Ayden kept crawling until his hands bumped into a vertical surface. He opened his eyes a tiny crack. It was William's desk. William himself was nowhere to be seen. As Ayden had hoped, the light was focused primarily on the settee area in which they had been sitting. At William's desk, it was still painfully bright but not such an assault. No doubt it had been installed as an essentially harmless defensive weapon designed to temporarily disorient an attacker. Combined with the virus, it had become a lethal offensive weapon.

Ayden dragged himself to his feet and studied the surface of the desk. A built-in touch screen displayed unfamiliar icons. He tried touching one of them. He could just make out the text against the glare, "biometrics not recognized." A small, black square appeared, probably requiring Hanford's fingerprint.

The desk had a column of drawers without handles running down its left-hand side. Ayden tried pushing on them, tried pulling them open from the edges. They didn't move.

Turning from the desk, he peered around the room. Wincing with every step, he crossed to the door and tried the handle, which moved freely but with no effect. He thumped the door, spearing pain through his skull. The door was cold, solid. It hadn't budged. It was likely reinforced and heavily secured.

Moving back across the room, Ayden hefted the chair onto the desk then gingerly climbed up beside it. The frosted glass ceiling panels were out of his reach but, lifting the chair over his head he could push at them. In most offices they would be simply sitting in place, easily lifted and dislodged. These didn't move at all. Ayden lowered the chair then swung it up as hard as he could into the ceiling. The chair bounced off the undamaged panel, crashing to the floor. Ayden collapsed onto the surface of the desk, weeping with pain.

He lay still for an age, then roused himself to push the chair over to a wall and try again against a wall screen, to no effect. He slid down the wall. He was in a highly secure, electronically isolated box. Floor, ceiling and walls were fortified and, no doubt, sound proof. There was no way out. Even sitting still, his head throbbed with agony.

Suddenly, he was plunged into absolute darkness.

He was still in great pain but the absence of light lifted a tremendous pressure.

He sobbed with relief.

Then, with a flash, the wall screens displayed huge images of Hanford's face. From every wall, he peered down at Ayden.

"I'm truly sorry, Ayden. I tried everything to make you my ally. I failed. We failed. Unfortunately, you are almost my match, too dangerous to leave alive. Don't bother trying to respond. The feed out of there right now is very low resolution video only. This transmission is heavily encrypted but I can't risk even the tiniest possibility of it being intercepted and decoded, of you saying something to reveal your location. I know the government has more powerful decryption resources than it lets on. Soon, William Hanford will also be gone from the world, but my work will continue.

"You've made me into a murderer and I hate you for that, but I also forgive you, so listen carefully because I'm going to try to spare you more pain. You've had the virus in your system for nearly ninety minutes. I estimate you'll fall unconscious within the next thirty minutes and be dead within an hour. I'm going make sure of that by assaulting you with light and sound again to accelerate the swelling. This will cause even more pain, but that's not why I'm doing it. I'm doing it to hasten your death, against the vanishingly small possibility of escape."

Ayden heard a tiny click from Hanford's desk.

Hanford continued. "I've just unlocked my desk. There's a syringe in there, in the top drawer, in a black case. It contains a very powerful tranquilizer. Spare yourself half an hour of the most acute agony you can imagine. Inject yourself. You don't have any alternative. Even I couldn't get out of that room, and I know everything about its construction. You have one minute to decide.

"Goodbye, Ayden."

Hanford's image disappeared. The wall screens provided a soft glow, enough for Ayden to see the desk, to use the syringe. Grimacing with pain, he considered it. It was tempting to take the easy way out.

Instead, he thought desperately. His prison must have some weakness. He staggered to the door to inspect it again.

It happened again. This time the light was room-wide, more intense.

Searing, blinding light pushed against his eyelids into his brain.

A shockwave of sound knocked him to the floor with palpable force.

Utter darkness and silence returned. Ayden lay on his side gasping in pain. His head pounding furiously.

Then another assault, light and sound.

Darkness, silence.

Light, sound.

Again. And again.

With each assault, the pain increased. He had never imagined anything could hurt so much. He was less than a body length to the desk. He inched toward it, desperate to do anything to stop the pain.

He would use the syringe.

He reached the desk. Light and sound smashed into him. He couldn't drag himself up to reach into the drawer. Darkness, silence. Light, sound. Ayden was swept into an ocean of agony and crushed. His body curled into itself.

He was aware only of pain and light and sound.

Darkness, silence.

Light, sound.

Finally, blissfully, the assaults receded. The universe receded, shrank to nothingness. Disappeared.

12:
APRIL 2034

The camera panned across the length of the train. The locomotive and the first few freight cars had derailed and concertinaed across the track. The rest of the train was intact and surrounded by hundreds of people and a growing collection of every type of vehicle imaginable from big battered trucks to donkeys and wheelbarrows.

People swarmed the freight cars like army ants and – like ants stripping a carcass bare – they unloaded the train's cargo at a wild pace.

The camera swung to show a man in the foreground, his face obscured by a dark grey cap and a dirty green handkerchief. He brandished an all black AK47 assault rifle with what looked like a grenade launcher slung under its barrel.

"I, Khalil Dahab, claim the contents of this train in the name of Africa's starving millions."

There was sporadic gunfire in the background and one of the railcars was on fire.

"For too long, our corrupt government has encouraged agricultural colonialism. For too long, our people have suffered – looking through fences at lush crops, watching ships head to China stuffed with our food while we starve."

NILE TV, "OUTLAW CLAIMS TO BE A MODERN-DAY, AFRICAN ROBIN HOOD," ENGLISH.NEWNILE.NEWS

Thursday, April 6th

Eons later, a voice.

Someone familiar. Soothing, talking quietly, calmly. He couldn't follow the words.

But the pain was still there. He slipped away again.

Saturday, April 8th

He woke to near total darkness. He didn't know where he was, how he had gotten there. He wasn't even sure who he was for a moment. He tried to look around. Pain returned. His head ached. His eyes ached. But the pain was nothing compared to what he'd experienced before. Memory returned, came flooding back.

He was only covered by a sheet and thin cotton blanket, but he was warm. A faintly antiseptic smell pervaded.

"Hello?" His voice came out in a croak.

A door opened briefly, throwing a harsh glare into the room.

"Ahh…"

"Sorry," a soft voice said. The vague shape of a woman approached him in the darkness. "The corridor is very dimly lit. I'll up your painkillers."

"Where am I?"

"Safe. Your family is here. They are going to pay you a quick visit, then you will need to rest again."

There was a quiet tap on the door. Ayden closed his eyes quickly.

"Come in," the nurse said quietly.

"Ayden? You're awake?" Cate said as she entered. Marc closed the door behind them.

"Hi, Mom. Hi, Dad." Ayden opened his eyes. "It's good to almost see you."

"Glad to see you awake, son," Marc said quietly.

"Oh, sweetheart, you had us so worried." Cate kissed him gently on the forehead. Her hand slipped into his as she sat beside him.

"Sorry. No choice. Had to find out."

"Don't worry about any of that now, my darling. Just rest. They say you have every chance of making a full recovery, but it was close Ayden. Too close. You very nearly died."

Ayden felt a tear splash onto the back of his hand.

"Your brain is still bruised, but you're going to be fine. You'll be up and about in no time causing trouble again." She sounded like she was trying to convince herself.

"They have to catch him. He's probably in a plane. They…"

"Shhh…you shush now, darling. Stop please. You've been in an induced coma for nine days. Hanford is dead. His plane went down in the mountains."

"No. He can't be. It's a mistake."

"They found his body. It was a charred corpse but they matched his DNA, dental records…everything."

"No. He must have faked it."

"Don't worry about that now. Just rest and get better. Believe me, there are lots of people investigating everything to do with this mess."

"Not yet. I've got to tell them. Let me tell you. I'm sure they're watching us. Listening."

There was another quiet knock on the door. Ayden closed his eyes against the glare as it opened.

"I'm sorry but he needs to rest—" the nurse said, turning.

"Five minutes," Henri said. "Leave us."

"Yes, ma'am. I'll be back in exactly five minutes. Ma'am."

Henri closed the door. "Yes, we're listening." She turned to Cate and Marc. "I suggest you leave us too. The more you know, the harder it will be for you to return to a normal life."

Cate didn't respond. She glared at Henri.

Marc gently took her arm. "Come on, sweetheart."

"Let me tell her, Mom. She's right. Later, we can talk about how much you want to know. Right now, I have information they need and it can't wait."

Cate hesitated.

"Mom, I risked my life for this information."

Cate let out a huge sigh. "Okay." She kissed him again then, turning to Henri, her voice chilled, became a hard-edged weapon. "I'll talk to you again later. Be quick. Let him rest."

When the door was closed and Ayden could reopen his eyes, he found Henri sitting close to the edge of his bed. She wore a suit and tie.

"She threatened to kill me, Ayden," she said smiling. "She said if you died, she'd kill me, or die trying. Whispered it into my ear."

"Wow." Ayden managed a small smile.

Henri's smile faded, replaced by a tired frown. "I'm glad you made it, Ayden." She shook herself. Her tone turned lighter, wry. "Quite aside from your mother, it would've been bad for my career if you'd turned up dead."

"I was beginning to think I'd made a serious error. That you weren't following me after all."

"We followed you alright, but you didn't make it easy. Nearly lost us on two separate occasions."

"Sorry. Had to make it look convincing in case Hanford was receiving reports."

"As he no doubt was."

"So what took you so long? It wasn't fun in there."

"We had no way of knowing what was going on, that anything was going wrong. Finally, we saw Hanford leaving but no sign of you. Then we discovered that the place was built like Fort Knox. They couldn't unlock the door. We had to blast the damn thing open. Took out most of the wall."

"Well, I'm glad I was unconscious for that. When I first snuck out of the cave, there was no doubt in my mind. You'd been sending me a message, dropping hints. You'd let me escape so I could find out what was going on. You'd keep tabs on me. But when I couldn't get out of that room, I wasn't so sure. Perhaps I had really just overheard a conversation about security minutiae."

"Sorry. I couldn't just open the door for you. It wasn't sanctioned."

"You didn't get permission to let me out?"

"No way. Gambled my career but," she smiled, "if you have decent intel for us then I'll get away with it. I didn't have explicit orders not to let you out and I did have orders to get results any which-way so, if it proves successful, they'll call it brilliant initiative."

"Why did you take the risk?"

"Why did you? I couldn't ask for permission, couldn't be sure Hanford didn't have people working for us. That's the official reason. Also, if I'd asked, they'd have said no. Now, we're running out of time. Tell me quickly what you have, before the nurse and your mother come back in here and throw me out."

"It's not much but it's important. Very. First, Hanford is behind it all, he's not just a player. He's the key figure. Second, it's deliberate, engineered, and he'd been planning it for over a decade. That means he will have made it intentionally difficult to vaccinate against, but he's careful, *he'll* have one. Third—most importantly—it's not random. It's targeted. 'Speciation within existing diversity,' he said."

"What?" Henri said, her head jerking back.

"He's trying to create a new sub-species of human being."

"That's not possible! Surely?"

"He's at least as smart as I am. He's been working on the idea since he was a teenager. He's had one of the biggest multinational biotechs in the world behind him. And he thinks it's possible. What do you think?"

"Who is he targeting? What sort of sub-species?"

"I couldn't get that out of him."

She stared at him.

There was a soft knock on the door. Henri went to it and Ayden heard a brief argument conducted in hushed tones. She returned to his bedside.

"I'm about to be kicked out, Ayden. Anything else?"

"One more thing. Mom told me that Hanford is dead." Ayden watched her closely.

"Yes. His corporate jet went down."

"Don't believe it. He said William Hanford would be gone from the world, but his work would continue. He could easily have faked his death." Ayden held up his hand as Henri started to protest. "They didn't find a whole body, did they? It would have been easy for him to set it up, to hack his DNA records. He'll have a new identity prepared. I'm sure of it."

She regarded him levelly. "I'll pass on your convictions."

"It's absolutely imperative we find him. He isn't dead, and he's trying to re-engineer humanity on the fly. It's wildly dangerous. We have to find out exactly what the virus selects for and we have to get hold of his vaccine."

The door opened. Ayden's eyes were becoming less sensitive to the light of the corridor. He kept his eyes open, watching Henri. The nurse had brought reinforcements, a doctor by the look of him. "Major Soulandt, I really must insist."

"Yes, yes. Coming," she said without looking away from Ayden. "See you soon. Rest up, there's a lot more to do. And… thank you."

Ayden relaxed, allowed the weariness to descend. He'd passed on the most vital information.

He slipped into sleep.

Friday, April 14th

For most of the week, Ayden had slept. Each day, he'd found slightly more energy, become slightly less sensitive to light and noise. Cate and Marc had visited daily, as had Tess.

His otherwise sterile hospital room had gradually accumulated character through Tess's flowers and the throw rug Cate draped across his bed, and the pile of printed-on-paper books and journals which Ayden had resorted to when the staff refused to give him any access to electronic communications.

He was at the Walter Reed Military Medical Center in Bethesda. Tess and his parents were staying at a motel nearby.

Each day he'd been visited by a man who introduced himself as Garry and who had instructions to very gently debrief Ayden, to "fill in the blanks." At the end of the week, Henri visited again.

"You can go in now, Colonel Soulandt," Ayden heard as the door opened.

"Hi, Henri. Did I just hear right? Are congratulations in order?"

Henri gave a brief laugh. "It was promote me or discharge me I suspect. I got lucky, I'm still useful. I have to say, you're looking better. Still skeletal but no longer on death's door."

"Thanks, I must be recovering. I'm bored beyond belief."

"They tell me you're making good medical progress. The bad news is you've been deemed sufficiently recovered for a more thorough debriefing."

"Oh, joy, and then?"

"That is indeed the question on everyone's minds...what to do with Ayden Walker."

"A conundrum clearly, but not as important as finding Hanford, or as responding to his virus. I can help find him if you haven't already." He examined her closely. "I can help you to deal with him and the virus. Help stop it, counter its effects."

She returned his gaze steadily, her face revealing nothing. "Yes, you could."

"But not locked in a hole in the ground."

"Ayden, you know they'll never let you out of their sights, don't you?"

"I do. You will watch every move I make...but since I know so much already, why not make use of me?"

"I agree." Ayden was sure she had placed the slightest emphasis on the "I."

"But leave my family out of it. Let them go home."

"That'll be my recommendation if they assure us of their cooperation. But now it's your turn for a warning. You're in pretty deep. You go any deeper and you've got no chance of extricating yourself or leading anything like a normal life. *And* you'll have to do as you're told."

"What chance do I really have of leading a normal life now?"

"Okay. I'll pass on your request and my recommendation."

"That's it?"

"Not by a long shot. I have no idea what they'll decide. In the meantime, it's more debriefings and exercise for you."

Wednesday, April 19th

Ayden was grinding his way through yet another physical training session when Henri walked in. He was in one of the small rooms adjoining the main rehabilitation gym in the Intrepid Center. Translucent panels set into a curved wall provided privacy and natural light without glare.

"Hi, Ayden. You're looking a lot more human," she said.

Ayden hopped off the step machine he had been working on. "Thanks. I tire quickly but I'm definitely on the mend." He grabbed a small towel and wiped the sweat from his face.

"Sorry to be so hard to reach. Been a little busy."

"I can imagine. I can only imagine because I'm being kept completely in the dark. Henri, you've got to get me out of here. I have to know what's going on. I won't be of much use completely insane. Please?"

"Sorry. You're status is in limbo. I tried to get you a phone but 'til the deputy director makes a decision, the orders are strictly zero communication with the outside world."

"Even that'd be better than nothing. Deputy director of what? Military intelligence? Homeland Security? NSA?"

"Can't say."

"How long is this decision likely to take?"

"No idea. Depends what else is going on. You know how it is. Difficult decisions without critical time horizons keep slipping to the bottom of the pile."

"Can you at least tell me if you know where Hanford is?"

"That's an easy one. There's no sign of him. Either he's really dead or he's doing a good job of hiding."

"Listen, Henri, time *is* a critical issue. If you haven't found him, you don't know what he's doing. But he'll be moving. Fast. Implementing contingency plans. Can you get me in to see this deputy director of yours? Can I meet with him?"

"Come on, Ayden! I've only met him myself a couple of times, and only to be grilled and dismissed."

"You can do it, Henri. Get me five minutes. I won't embarrass you, I promise. He'll thank you for arranging it. I just need five minutes to make my case."

Colonel Henri Soulandt sighed. "They're right. I do have a weakness when it comes to you. I'll try. No guarantees."

"Thanks, Henri."

Friday, April 21st

"Thank you for agreeing to see me, Mr. Deputy Director."

They sat opposite each other at the cozy end of the deputy director's office, in which traditional overstuffed chairs clustered, surrounded by wood paneled walls and oil paintings in ornate frames. Back to back with one of these armchairs, a lone chair faced the other half of the room, which was otherwise empty and appeared to be a platform overlooking snow covered mountains. The walls and ceiling, and probably even the floor, were edge to edge screens which would provide seamless peripheral vision support to the deputy director's specs in full virtual reality mode.

"You have three minutes, Walker. You've just wasted nearly 5% of that time."

Henri had advised him not to try to convince the deputy director with rhetoric or moral suasion.

"You won't change his priorities, Ayden," she'd said. "Your only chance is to convince him that you can assist him to achieve his goals, and you're going to have to be painfully concise."

The deputy director's goals were clearly focused on the well-being of the United States of America, which usually meant the well-being of the government and its security apparatus.

"Hanford's not dead, and I think you know that or are concerned that it may be the case. Know also that he's at least as intelligent as I am. Even in hiding, he has access to significant resources, and he has his own agenda and will follow it ruthlessly. His ambitions are global and he doesn't give a damn for national security as you see it."

Ayden paused. The deputy director didn't respond.

"Finding him is urgent not only because he will have the vaccine your scientists are failing to derive but also because we don't know what he meant by 'speciation within existing diversity' and, even worse, he will have made contingency plans which can be implemented quickly. I am the best person to find him, but you wonder whether it's worth the risk."

Ayden paused again. He was getting nowhere. The deputy director's eyes had started to stray. He wasn't interested. Capable, well resourced enemies with global ambitions were this man's daily fare. He wasn't convinced he needed Ayden.

Three minutes or otherwise, Ayden figured he had about another 30 seconds before he was thrown out. He took a deep breath and plunged in. He had wanted to avoid this.

"But here's something you don't know. Something I have kept from Colonel Soulandt and from all your debriefers. Even my parents don't know the full extent of it. Time is on his side. Barring misfortune, he'll outlast you and your replacement and his replacement. Hanford can play a long game which you can't hope to match."

The deputy director's eyes riveted on Ayden's. He froze. Now Ayden had his full attention.

"We weren't just modified for intelligence. We were also engineered for longevity. Hanford and I may live for in excess of a 150 years. We're only in our twenties now, sir. He could be a factor this office needs to contend with for the next hundred years plus."

The deputy director's gaze was fierce. Ayden had thrust him onto an unfamiliar playing field. He'd also held up the promise of a glittering prize.

Now he'd find out how the dice would fall. Henri had told Ayden she believed the deputy director was both highly intelligent and essentially a good man, albeit one forced to adopt a ruthless philosophy. Ayden hoped Henri was right. A short sighted man might have Ayden killed. A small minded man might try to lock him away for the rest of his long life, for military scientists to study.

But Ayden had also demonstrated that he could maintain big secrets despite expert cross-examination in successive debriefing sessions.

The deputy director leaned forward. "You can support this claim?"

"I can describe some of the adjustments which Hanford described to me. Your staff will, no doubt, already have analyzed my DNA. They'll be able to confirm my unusually long telomeres and over time, they'll be able to track my low rate of replication error. For the rest of the detail, we need to find Hanford."

"How convenient." The deputy director smiled. "And you agree to maintain this secret."

"I have the most to lose should this become public. I am very much hoping you will keep it secret for as long as possible."

The deputy director leaned back and closed his eyes. There was a soft tap on the door. The deputy director touched an icon on his desk. "Clear my schedule for the next hour." He didn't wait for a response. There was something predatory about his smile. "Okay, Walker, what do you want?"

Saturday, April 22nd

"How'd you do it, Ayden?" Henri said. She was now seconded to the Office of the Director of National Intelligence and reported solely to the Deputy Director of Asymmetric Threat Assessment. Her objective was to find Hanford, and to make maximum use of Ayden to do so. The deputy director had also made it clear to her that her career was over if she let Ayden run loose again without his explicit permission.

They sat in a small Pentagon meeting room at a bamboo-board table surrounded by frosted glass.

"Sorry, Henri. Can't tell you." Ayden laughed.

"'Okay, wise guy. I'll find out eventually. At some point, I'll 'need to know.' In the meantime, I'm used to it. What deal did you manage to do for your parents? And Tess?"

"They are going home. They're out of it. They'll have to sign yet more legal documents and there'll continue to be surveillance but they'll be able to lead relatively normal lives. Mom will be reinstated to the bee implementation. George Reyes, too. It's really just an overseeing role now, but it'll help to remove the black mark from their reputations. They'll even get a chance to help Ramy Da Silva; they'll need someone in the Asian time zone."

"Assuming they agree."

"They're smart enough to know I'm going to stay caught up in all this no matter what. Besides, it's not like I'll never see them again. I just won't be allowed to tell them what I'm doing and all my contact with them will be subject to close surveillance."

"Your parents will be okay with that?"

"They'll live with it. They didn't really want any of this. They've just been supporting me. They went to a great deal of effort to give themselves – and me – a quiet, scholarly existence. It's time for me to allow them to go back to that."

"Can't see Tess liking it."

"No. But it's not like any of us really have a choice."

"No."

"So I have to convince her to leave all this behind and return to a normal life. I owe it to her. I'm effectively going to be under a sort of house arrest for some unspecified but extended time period and I can't let that happen to her. Now, I have to get her out. Even so, I hope I haven't made a complete mess of her life."

"She's a big girl, Ayden. She'll recover."

. . .

They were in his hospital room, at Walter Reed, sitting on his bed.

"You were lying to me the whole time, weren't you, Ayden?"

"What? C'mon Tess—"

"Did you deliberately set out to give me the impression that you were thinking of siding with Hanford, while really you were trying to play him?"

"No. Well, not really. I guess I did emphasize one side of the argument, knowing he was watching us, knowing he'd watch your reaction but—"

"So that's a 'yes' then. You deliberately misled me, you lied to me to fool Hanford."

"Tess, it's not as simple as that. It's not that black and white."

"You're still not prepared to be honest about it!"

"I am. Okay, you want the unvarnished truth? Yes, I was conscious of how you and my parents would interpret my comments, but it wasn't a coldly calculated Machiavellian plan, Tess. I was winging it, flying by the seat of my pants. Still am. I don't know how all this is going to pan out. I can't tell you how long it'll take for me to reclaim any sort of normality in my life. I just don't know."

"Why can't you just walk away? You haven't done anything wrong."

"You know it's not that simple, Tess."

"Why not? What would they do?"

"I don't think I want to find out."

"We could go to the press. Tell them everything. The bastards wouldn't dare touch us then."

"And I get to spend the rest of my life as a one-man freak show, hiding from all the loonies trying to buck the odds by putting a bullet in my brain. But that's not the point anyway. What Hanford's trying to do is breathtakingly dangerous and I wouldn't walk away if I could."

"And I don't figure in your calculations at all, do I? We don't, despite how we feel about each other." She paused abruptly. "Actually, I'm a complete idiot, I've only just realized that I haven't the faintest idea how you really feel about me. You're the most frustrating man I've ever met, but I love you, Ayden, and that's all I need to know to set my priorities, but it's different with you, isn't it? How do you really feel about me, Ayden? Do you feel anything at all?"

"I'm sorry, Tess, but love doesn't conquer all. Maybe it's what makes it all worthwhile, but it doesn't—"

She leapt off the bed and spun to glare at him. "Don't you dare patronize me! Unlike my mother, I realized long ago that Hollywood love is no more than a pretty fantasy. I know that in the real world love causes as many problems as it solves. This conversation is about priorities and I'm just not as important to you as finding Hanford. Am I?"

They stared at each other for an age.

"No. My happiness is not as important as stopping Hanford's science experiment with humanity and nor is yours. I knew I was risking my life, Tess. I didn't do it lightly."

"And in the meantime, your plan is to shut me out."

"Tess, I don't have any choice, and neither do you. I'll work out some way to loosen the noose around my neck, but…it could take a very long time. You have to do this, Tess, for your sake and for mine."

"And just how long do you expect me to wait for you?"

He paused, then forced himself to say it. "I don't expect you to wait for me."

"You don't–" Her fury ascended to a new level. "I won't sit around twiddling my thumbs waiting for you to get bored playing secret agent, or worse, playing God."

"Good."

"Good! That's it? Well, I guess that answers my question."

"Please, Tess, you know—"

"Don't you dare ask me to accept this gracefully and don't tell me what I know. I'll tell you what I don't know. I don't know you. I thought I did. I thought you were the most considerate, wonderful man in the world, as well as the smartest. But it turns out you're like all the rest. I'm allowed to bask in your sunshine while it suits you but as soon as things start getting messy, I'm out of there."

"Tess—"

"No. You go to hell."

Tess walked away. Out of his room, out of the building, out of his life. He lay back on the bed and stared at the ceiling for a long time.

W.R.R. MUNRO

13:
MAY 2034

The big stern trawler swayed in the waves. There were at least a dozen men down on the fishing deck, shouting to each other over the noise of the winches, which strained to haul the taut net on board. A great mass emerged from the ocean but the rear doors were left open and most of the men turned away from the net, disgust evident on their faces and in their movements. One of the men worked at the back of the net, releasing it, spilling the gelatinous mess back into the ocean. Another man climbed toward the camera and shouted at it.

"Jellyfish! Fours days out and it's just fookin' jellyfish again! How we supposed to make a fookin' livin'?"

LADY GRACE WEBLOG, "THE NORTH ATLANTIC'S FISHERIES HAVE COMPLETELY COLLAPSED. SCIENTISTS BLAME OVER-FISHING, WARMER OCEAN TEMPERATURES AND POLLUTION," ATLANTICFISHERIES.INT

Tuesday, May 9th

Ayden was inscape in an office adjoining Henri's. She'd commandeered the offices on base at Walter Reed though she spent a lot of time elsewhere, presumably with the analysts somewhere in Virginia or D.C.

He spent as little time in Terry O'Brien's personal scape as possible. Looking around, he couldn't shake the feeling of wrongness. It wasn't simply that it was foreign. He'd had to get used to mil-spec gear with its bland scapes and hierarchical association rules, to not having full admin rights. But this was different. This was as individual and quirky as his own space. As personal. It felt like he was peering into someone else's mind.

O'Brien was on their shortlist because he was a wealthy recluse with an obsession for longevity research and a passing resemblance to Hanford, though he was apparently in his mid forties. He lived in a private estate on what was almost an island in Lake Volta, in Ghana. Africa. A narrow land-bridge provided the only access to the estate. O'Brien had inherited a cocoa estate but left its running in the hands of the local Ewe people who, in turn, fiercely protected his privacy. He had almost no face-to-face contact with the outside world but was an active participant in longevity conferences and invested substantial sums in research, which didn't appear to have borne fruit. Either that or he wasn't sharing the results.

They had trouble getting hold of a recent photo but they'd assembled a massive dossier and finally managed to hack into his gear. O'Brien's estate was sufficiently remote that it accessed the grid by satellite and Henri's team had piggybacked a software update, giving Ayden unfettered access to O'Brien's personal scape.

"It's not him," he said to Henri's image, windowed in his specs. "Not if the scape's real."

"That was quick."

"I'm sure of it."

"Your reasoning? Or do we just close the file with 'Ayden's sure?'" She softened it with a slight smile.

"That'd work for me."

"Mmmm."

"Firstly, because, while he may not have contact with the outside world, he certainly has lots of contact with the locals. He's a "white stool" which is some sort of honorary but prominent position and, well, it looks like he may have children with several of the women who look after his household."

"Unless it's all planted. Infiltrating the village would be a long-term job if it was doable at all, let alone the household, so we can't confirm the scape is telling us the truth," Henri said.

"Could the whole scape be fake, an artifact against the possibility of intrusion?"

"We don't think so. Whoever he really is, he spends most of his waking hours inscape, even when interacting with his household."

"Tell me about it." O'Brien wore his gear during sex, recorded everything. Left the files all over his scape. "Which means O'Brien isn't Hanford. Besides, that scape's a mess. It's chaos. I can't believe Hanford would spend any time in there."

"You realize he's the last one?"

"Did you really think it'd be so easy?"

"Easy? We've looked at every person involved with commercial organizations or research vehicles linked to Genenco. We looked at people with no apparent connection to Genenco but who're associated with bio-medical research and at people who've aligned themselves with Genenco or Hanford causes. Our initial list ran to over thirteen hundred persons of interest."

"That does worry me a bit, Henri. Your team only gave me the shortlist of two hundred or so names. They crossed the first eleven hundred off pretty quickly."

"Give us some credit, Ayden. We've done this before. I've got a large team of experienced analysts working in shifts, twenty-four by seven, with access to a serious amount of information."

Ayden knew better than to ask for more detail about the team. "Need to know" was a tedious tyranny but he'd given up fighting it. Henri's team of "facilitators" had an acute understanding of the information, personnel, and computational resources available within the disparate collection of organizations that constituted the intelligence community. More importantly, they knew how to access and marshal those resources. Even with a level one priority from the deputy director, they still needed to be adept office politicians as well as analysts. Whoever they were, the team had amassed a horrifying amount of information about everyone on their shortlist and presented Ayden with the dossiers they thought warranted his direct attention.

The "short short" list had contained twenty-nine names. These men were targeted for even more detailed study. Friends and associates were very discreetly questioned. Ayden actually sighted most of them, virtually, via spec "walk pasts" in public places.

And he'd just crossed the last one off the list.

Problem was, Ayden was being kept in the background. The facilitation team knew they had a "witness" who was providing personal information about their target but all contact was via Henri. Apparently, it was standard practice. They would do all the investigation. Ayden was to focus on developing the profile for the man they were looking for and review their findings.

"Let's go over it again."

Ayden didn't quite manage to suppress his groan.

Henri continued, relentless. "Hanford will have long planned for this contingency so the identity will be well established. He'll be very low key. A background figure. Easy to overlook. He won't appear overly powerful or influential but he will be pivotal. He'll crop up repeatedly when relevant organizations take action. His real age is mid twenties but he may appear significantly older. He won't necessarily be a scientist but he'll be in a position to influence scientific research. He'll have been extremely reclusive but he may not present himself that way. His colleagues will rarely, if ever, have met him in the flesh but he'll have talked about social occasions and face to face business meetings. The clincher will be that it'll all be anecdotal rather than first hand."

He just looked at her.

"Okay. Take the rest of the evening off. Get some exercise or something. Make sure you're fresh for tomorrow."

"Henri," he paused, considering how to say it.

"Spit it out, Ayden."

"This isn't working."

"Then tell me what to do. How to go about it differently. Cause till you do, I'm going to stick with dumb persistence."

"I've produced a shortlist of my own, but you know that. You watch everything I do inscape."

"Your collection of women and old men and others way too high profile or whatever. All stateside. All very connected. All very clearly *not* Hanford. I've been wondering when you'd be ready to talk about them."

"It's just another layer. We need to look for the person Hanford's hiding behind, a real person whose strings he's pulling. His one weakness is he won't be willing to just hide. He'll want to continue his work."

"Okay, well, I'm not at all sure about this but I'll get the team on it."

"No."

"No?"

The whole process was too indirect. Ayden couldn't pick up on subtle clues working through a massive team of agents. He needed to personally meet the targets, interact with them.

It wouldn't be easy convincing the deputy director.

• • •

"Truth is, I'm not really sure how it'll work," he told the deputy director's avatar which appeared to sit opposite him, in the middle of the room. His office was fitted with multiple cams so the deputy director could see Ayden in high-resolution 3D but there was no telling where the deputy director was or what he was doing. His avatar was, no doubt, deliberately inexpressive.

"Clearly none of these people is Hanford, even with the cleverest disguise or cosmetic surgery. Equally, they all have numerous connections with endeavors which Hanford was interested in."

"So, we keep investigating them, accepted," the deputy director said. "That doesn't require you to interact with them directly. We can send in agents. It's what they do and they're good at it."

"Problem is, I won't know what I'm looking for 'til I find it. When I do find it, I'll recognize it but in the meantime, I just have to go fishing."

"Right now, young man, your official status is critically ill, though of course, we've had to bury any mention of DARPA's meningitis, as Hanford knew we would. You could succumb to your illness at any time and Hanford, assuming he's actually alive, can't know for sure what your real status is or whether or not you're helping us. Convince me I shouldn't keep it that way."

That the deputy director's words could be interpreted as a threat wasn't lost on Ayden.

"Our problem is time. Let me guess, your scientists are continually telling you that a vaccine is just around the corner despite a few unexpected hurdles, right?"

The deputy director didn't respond.

"What they're not taking into account is that he will have deliberately engineered it to be hard to derive a vaccine for using all the standard methods. If you want a vaccine, you're going to have to find Hanford. And we need that vaccine! It's not just about the population level. He's trying to create a new breed of human, and we don't even know the characteristics he's targeting."

"I'm acutely aware of the urgency, Walker, but at the moment there are very few people who know that we don't necessarily accept Hanford's death, that we're actively looking for him. All our investigations to date could simply be us trying to work out what had been going on, who's behind the bee virus, if anyone. Your sudden reappearance would be tantamount to waving a flag and shouting that we know he's alive and we're coming after him, so all he has to do is pull his head in and he'll really disappear."

"Unless we muddy the waters with a convincing enough cover story, and I interact broadly enough. He won't be sure what's going on. He'll be concerned but not, hopefully, enough to warrant dumping his identity. We have to assume he hasn't many backup identities in place. It would've cost too much of his own time to maintain them. They'll be invaluable to him."

"What cover story do you propose?"

"We show him the government making an unofficial but concerted effort to rehabilitate Ayden Walker's reputation and career. You employ a PR agent to my case. You appoint me to various government advisory committees and research funding bodies.

That activity provides me with the perfect explanation for suddenly meeting with a broad group of people who operate in the same arena."

"And what would be the unofficial explanation for the government wanting to thrust you into a position of prominence, particularly if we think Hanford is dead?"

"It would be vehemently denied, of course, but it would be rumored to be a payoff."

"A payoff. For…?"

"For me having provided information of value or for my silence in some regard or both. As it would be denied, no one would know for sure, which would make the rumors and speculation all the more enticing. Hanford will assume I've done a deal with you, handed over the data disk perhaps."

"Interesting proposition, young man. It's sufficiently close to the truth that any insider with an inkling of your activities and whereabouts would find it credible. It could work."

"Thank you, sir. The reason I asked for this meeting to be just you and I is so that, even internally, we could allow people to assume it is some sort of payoff… in case Hanford has someone close to your office."

"Hmmm…you have a devious enough mind to be good at this game, Walker. What really strikes me though, is how convenient this particular solution is for you." The deputy director's eyes drilled into Ayden's. "We may find Hanford. That's a possibility… assuming he's alive. In the meantime, what we would definitely be doing is strengthening your position, making it harder for me to keep a lid on you."

The deputy director paused, waiting for a response. Ayden said nothing. The silence stretched. The deputy director waited. Eventually he smiled. "It's amazing how few people can resist responding through a silence that long, Walker, but then I'm learning not to underestimate you. However," the smile disappeared, "I require a response."

"You still have the secret of my longevity to hold over me. I really don't want that made public. I made myself even more vulnerable to you than I was already when I revealed that."

"Now you want me to reduce some of that vulnerability. I could say no, do it my way, send in agents, probe associates."

"But you don't believe that would be as likely to yield results."

"Hmmm."

"But you haven't decided whether it's worth strengthening my hand to maximize your chances of capturing Hanford."

"It's lucky for you I don't believe in mind reading, young man. I'll let you know."

The deputy director's avatar disappeared without another word.

Wednesday, May 10th

Ayden flicked through personnel dossiers. With a soft chime, an icon appeared, Henri's.

[Accept.]

[Ayden, do you have a minute?]

[Sure, Henri.]

A few moments later there was a soft tap on the door.

"Come in." Ayden had always been struck with Henri's politeness. She could have simply appeared without warning in Ayden's scape as the deputy director had, or physically walked into his office unannounced, but she used her rank softly. Ayden had come to regard it as a sign of grace. She sat and looked across the desk. Ayden noticed an amused smile on her face.

"What have you been up to?" Henri asked.

"Hopewell Foundation."

"The think tank?"

"Yeah, they've got a charitable trust arm which funds research into public health and education. I'm looking at key staff, some of the dossiers are a little thin though."

"Spoken like a true desk analyst. Flag any of interest and we'll double-F them."

"Double-F? Sounds unpleasant."

"Fatten their files."

'To be frank, I'm glad we don't know everything about everyone."

"We're working on it."

There was a pause as they regarded each other.

"I'm going to have to be very careful with you, Ayden Walker. One day, I'm going to wake up to find you've suddenly become my boss."

"I take it the deputy director said yes?"

"He did. After chewing me out because I couldn't come up with a better plan for finding Hanford. One which didn't require letting you out of the bag."

"Sorry if I'm causing trouble."

"Sure you are." She examined him. "Maybe you really are. Well, forget it. You learn to be thick-skinned in this job. Besides, he only did it for the sake of form. He couldn't think of a better idea either." Henri's smile faded. "He wanted me to make two things clear. Firstly, you will be watched very, very closely." Henri paused and locked eyes with Ayden.

"I understand. Zero privacy. I'm not delighted, but I accept it as unavoidable. At least for the time being."

"Second, you'll be on a very short leash. You have a habit of just going off and doing things on your own. According to your file, you did it when you were working for Reyes and you did it again, albeit with my tacit approval, when you broke out of the cave. But it can't happen again, Ayden."

"I understand."

"I'm sure you do but I'm going to tell you anyway. You need to be very, very cautious. If you do anything unexpected, the deputy director will…" Henri paused. Any conversation in Ayden's presence would, of course, be recorded. "I'm not sure what he'd do exactly but I'm pretty sure you wouldn't like it."

"I understand—" Ayden stopped as Henri held up a hand. She wasn't finished. A good soldier, she would deliver her boss's message in full.

"In practice, that means you run everything past me. Everything. You don't contemplate the slightest departure from the script without explicit approval. Got it?"

"Yes, ma'am."

Henri smiled. "*That's* what I've been waiting for. Now, let's see how much tax payers' money we can blow restoring you to your former glory."

Monday, May 15th

They'd appointed a legal team and a PR agency. The legal team set about clearing Ayden's name regarding the Hunt River case. Apparently, it was helpful, though not vital, that he had actually been innocent of any wrongdoing.

The PR agent was a man by the name of Mathew Rawlins. Rawlins was a tall man with steel grey hair. His conservative, charcoal-grey suit and white shirt never changed. Only his tie varied and it was always wide stripes and strong colors. Everything about him suggested consistency and reliability, as it was designed to.

According to his NSA file, he was born Reginald Lewis and had been of only average height until he underwent the lengthy and painful series of operations which added three inches to his height. He was thirty-eight years of age but his always serious demeanor and grey hair gave the impression of a man somewhere in his forties. Ayden wondered if the grey hair was natural. He suspected not. Rawlins was a product of his own creation.

Ayden had spent the entire weekend with Rawlins, including breakfast, lunch, and dinner. He'd quizzed Ayden about every aspect of the situation but was unfazed at how little Ayden was permitted to disclose. He'd also asked Ayden lots of questions about himself.

Now, on Monday evening, Henri joined them in the plush, private dining room they'd been monopolizing.

"This's an entirely straightforward job," Rawlins said. "You're being placed on the Board of the Hopewell Foundation and joining the EPA's Advisory Committee so that takes care of association with distinguished individuals and organizations. We should add a charity to the list though. Doesn't really matter what it's for, as long as it's well established, with prominent people on its board. Any preferences?"

"What about the Lew Institute?" Henri said. "Very high profile and full of movers and shakers."

"Okay, but what do they actually do?" Ayden said.

"Primarily medical research. Particularly where they perceive the pharmas and biotechs are neglecting a possible therapy because they don't forecast enough profit from it."

"Sounds like a cause I'd be happy to support."

"Well, that was easy," Rawlins said. "I'll check their profile but I am vaguely aware of them already…I'm sure they'll do. That does, however, bring us to the one aspect of this job which concerns me just a little…your personal presentation." He paused, awaiting a response.

Henri smiled. Ayden raised his eyebrows, inviting Rawlins to continue.

"I have, of course, been observing you closely over the last two days. The key here is for me to work with the material at hand. Most people are unconvincing when they try to project a completely false persona. The trick is to project the real part of you that's most appropriate, downplay those aspects of who you are which would be unsuitable. Now, generally speaking, your presentation is good. You come across as honest and intelligent. You are at your most impressive when focused on an issue that's complex and important to you. When you speak to such an issue, your clarity and precision is remarkable. When you are focused on such an issue but not speaking, you have an amazingly calm… stillness, I think it is. It lends you an air of inscrutability, which is excellent for being taken seriously." Rawlins paused again and frowned. "Then you go and ruin it all. You smile."

Ayden pointedly ignored Henri's guffaw.

"Pardon?"

"You smile. Some smiles, mine for example, maintain a reserved air. Others, like that of the Colonel here, contain some mirth but retain an edge, an aggressive flavor."

Henri's smile vanished. She looked taken aback but said nothing.

Rawlins continued, "Yours is too open and inviting, too warm. It makes you seem too nice. Nice equates with harmless. Powerless. Inconsequential. I gather from Henri here that, over time, people learn not to underestimate you, but we want to move faster than that. You will be meeting lots of people only briefly so first impressions are important."

"What do you suggest I do about it? I doubt I can convincingly change my smile."

"Don't smile. It's that simple…at least with men. I'm sure your smile works wonders with the fairer sex. One-on-one with women, you can smile—they'll interpret it as interest—but not with men."

"I have to admit, Mathew, I'm struggling to take this seriously."

"I know, but I'm very good at what I do and I'm telling you how it is. If you want to be taken seriously by powerful men, stop smiling. If you want to be underestimated, which can be useful in certain circumstances, beam away. Try it. Observe how people respond. You're a scientist. Experiment."

"I will."

"Please do. Now, Henri, you'll no doubt be supplying Ayden with a PA?"

"Wait," Ayden said. "I can organize myself."

"Not the point." He turned back to Henri. "Should be a male, no taller than Ayden. Imposing but not too beefy looking. Obviously intelligent. Most importantly, someone able to display convincing subservience to Ayden." He turned back to Ayden. "Establishes you in an alpha context. Next, D.C. office space…"

Wednesday, May 24th

Ayden's ridiculously large office screamed "designer." The walls and custom cabinetry were in muted chartreuse or high gloss grey with just enough burnt orange to add drama. When he'd shown his parents though his spec cams, Marc just laughed. It occupied the corner of the building and two walls were floor-to-ceiling glass that could be frosted at the flick of an icon.

He'd checked that they were real windows rather than high definition wall screens.

[Voice message from George Reyes.]

[Play.]

"Welcome back, Ayden. I was delighted to hear you were back on the merry-go-round, but on reflection not so surprised. I understand you're golden again and doing lunch all over the place so drop me a line. I plan to be in D.C. early next week, and I'd love to catch up."

[Reply. Voice message.] "Hi George, it's amazing what a good hairdresser can do. I've insisted they leave me a couple of meals free each week. Does either Monday dinner or Wednesday lunch work for you?"]

In truth, Ayden didn't mind the full schedule. It allowed him to immerse himself in the world of biotech movers and shakers, which would be vital if he was to sniff out Hanford. He did experiment with Rawlins' advice and it was a little sadly that Ayden concluded Rawlins was correct. "Nice" wasn't taken seriously. He found that by focusing intently on his mission, by studying the people he met rather than naturally interacting with them on a personal level, he tended to smile less and presented a more serious demeanor. On a personal level, however, he found these interactions with people less satisfying.

Ayden's earpiece chimed with the notes he'd assigned to the "personal assistant" minder, James, who occupied the outer office and shadowed Ayden's every move.

[Yes, James?]

[Sir, I have a call for you. Claims it's personal, but he's not on my list.]

Ayden sighed. Henri insisted that James actually do the job of PA even though it was nowhere near as efficient as just letting Ayden configure his phone properly.

[Who from, James?]

[Greg Fanshaw, sir.]

Ayden hesitated a moment. Anyone he spoke with would be investigated, but he was sure Greg didn't have anything to hide and probably they'd check him out just for trying to contact Ayden.

[Put him through, please.]

[Yes, sir.]

"Hi, Greg."

"Ayden? I didn't think I was going to get through. You're not an easy feller to get hold of."

"Sorry about that, Greg. Is everything okay? How's the crop?"

"It's more than okay, Ayden. Look, I know you're really busy and I won't take up your time, but I was determined to thank you. The bees really did their job this year and it's as if the trees are celebrating. At this rate, I'm looking to get a record harvest and I've got you to thank."

"Well, thank you, Greg, though it was a team effort and really we were only doing our jobs."

"I downloaded your EPA presentation; you weren't just a simple field researcher. And you gave a damn. I've heard you're in D.C. now. If you decide you want a break from the sharks, I want you to know you're always welcome here, Ayden, any time, we'd be pleased to see you."

"Thanks, Greg, I'll be sure to take you up on it."

"Good, and if there's ever anything I can do for you, don't hesitate one second, okay?"

They signed off and Ayden sat back in his ergonomically perfect chair. He *had* been just a simple field researcher though. Just one year previously. He'd love to take up Greg's offer, perhaps even invite Tess to join him.

To at least pretend to turn the clock back, for a day or two.

He missed Tess.

Ayden shook his head. He had work to do.

Friday, May 26th

The Board of the Lew Institute met monthly to review progress and discuss new funding proposals.

This was Ayden's first meeting. He remained silent throughout, observing. Having read the briefing papers, he paid limited attention to the proposals themselves, instead focusing on his fellow Board members.

It was a painfully dry affair. Their charter required transparency, so all formal meetings were recorded and could be accessed through the grid. As such, the meeting consisted mainly of a staff member reading a précis of the application and providing a very concise summary of the arguments for and against it. They sat around a huge, oval table in a generic boardroom, flicked through virtual papers and listened. Few questions were asked.

Ayden couldn't help but be disappointed. They were reviewing applications for bio-medical research, work that had the potential to alleviate suffering and save lives. He had hoped for impassioned argument.

The meeting finished without any new projects being approved.

As soon as the formal session closed, Juliette Otani, the Chairperson, approached Ayden with a small smile.

"Disappointed?" she said.

"It wasn't the fierce debate I thought it would be."

Her smile broadened. "Indeed. Would you care to join us for refreshments? Half an hour, Boardroom C?"

"Sure. Thanks."

Away from the public glare, Boardroom C had a bar set up at one end and the large board table was set for a formal meal. When Ayden entered the room, he found a few of his fellow board members comfortably sipping drinks near the bar. Over the next fifteen minutes, the remainder of the Board joined them.

Ayden smiled. Debate raged unchecked. It continued over dinner. This was where the real arguments were offered and countered, where the decisions were really made. Ayden's faith in the Lew Institute was restored. These people were intelligent, focused and passionate about maximizing the potential benefits of the funds they shepherded.

This was also where Ayden first heard of "she who." They were discussing a proposal to explore the use of a mitochondrial RNA approach to reduce DNA replication errors.

"You have to admit, Adam," Otani said. "We will be slammed if we fund this one. They'll say we should be funding medical advancements for the ill rather than life extension techniques for the wealthy."

"Oh, please, its anti cancer value is just as great as its anti aging value. Besides, it's a red herring to say only the wealthy will be able to afford it. Initially, it'll be very expensive, but eventually it'll get cheaper and the technology will flow on to everyone, as always. Besides, its main effect, if it works, will be to improve quality of life for the aged rather than significantly prolong life-spans."

"Yes, as much as everyone could do for themselves, at no cost, just by watching what they eat and exercising."

"Come on, Juliette, if we are to exclude self inflicted maladies that argues against probably half of all medical effort. I don't see why this research would be any harder to justify than a lot of the other stuff we fund."

"Let's be clear here. I won't hesitate to fight tooth and nail to preserve the independence of this organization. If it's really that important, we'll hunker down and take the flack. But I am wondering if this project is important enough to justify the heat we would get over it. What I'm trying to tell you, Adam, is that this one will meet serious opposition…"

A distinct pause floated through the conversation.

"She who?" Konrad Lee said. Until Ayden had joined, he'd been the youngest and newest board member.

No one answered and the pause settled itself until Adam Babcock admitted defeat. "Damn. Okay, it's probably not worth the pain. I'll let it go."

Ayden felt his pulse quicken. He pretended to concentrate on his dessert.

Tuesday, May 30th

Ayden forced himself to wait until after the weekend to invite Konrad Lee to lunch. They met in a typical Seattle businessman's restaurant in a discreet booth under a caricature of President Gore holding the world over his head, Atlas-style.

"Thanks for agreeing to join me for lunch, Konrad," Ayden said after they had exchanged pleasantries and ordered. "Actually, I was hoping to pick your brain a little."

"Ha, fair enough. Should warn you though, I'm still learning the ropes myself."

"Well, the meeting wasn't at all what I expected. There's obviously an unspoken protocol for the meetings and for the following dinner, and I was hoping you'd have some tips for the new guy."

Konrad laughed. "Sure do. Main thing, don't say anything in the formal meeting that you wouldn't want some journo with an axe to grind taking out of context. At the same time, it is an opportunity to get stuff onto the record, cover yourself…"

Ayden waited until they were sipping on coffee at the end of the meal to ask, "say, what happened with Babcock's RNA application? Who on earth is 'she who?'"

"Ah, yes. "She who must be obeyed," Sandra Allen, billionaire, patron of the arts, patron of just about every major charity, fierce lobbyist for or against dozens of causes. Knows absolutely everyone worth knowing and if she really gets behind something…or gets set against it, you cross her at your peril. Do it too often and she'll make your life hell. She's a true philanthropist though, does wonderful things for all sorts of needy groups. Even funds research herself, if she can't get organizations like us to do it. An angel with a howitzer tucked up her sleeve."

• • •

Within an hour, Ayden was back in his office with access to everything there was to know about Sandra Allen. The NSA had a six gigabyte file on her. Even the executive summary ran to ten thousand words.

She seemed way too high profile to be a Hanford crony, but was otherwise a perfect candidate. Ayden scanned the summary and watched hours of footage of her shaking hands with presidents, prime ministers, sheiks, and kings, dining with business leaders and making passionate speeches, pleading on behalf of African child soldiers, the long term unemployed in the U.S. and green tree frogs in Brazil. Her arguments were a mix of emotion and sentiment with hard-nosed pragmatism. She certainly couldn't be written off as a bored socialite. Her passion was real and she knew her subject matter. Overall, anyone would have to conclude both that she was a force to be reckoned with and that it was a good thing people like Sandra Allen existed.

Ayden pondered how to investigate further without setting off alarm bells? If Hanford was behind her, he would be on the lookout. Quite possibly, some of the team's earlier efforts may have already been noticed.

There were no obvious legal connections to known Hanford companies or trusts. Some of Mrs. Allen's crusades were even against Genenco activities. Ayden tried following the money, but once again he found no obvious connections to Gardner's estate or to known Hanford entities. She had inherited the money when her husband died. His money had come from supplying service stations nationwide with batteries for their electric car battery exchange services, which allowed people to simply switch batteries rather than wait for them to recharge, then when fuel cells became commercially viable, he'd deftly switched to distributing the hydrogen.

After he died, she sold the business for over six hundred million dollars, then defied expectations. She became an active investor in the twenties biotech boom and tripled her money in less than ten years. She engaged hordes of investment advisers, accountants and lawyers to manage the day-to-day aspects of her affairs but she clearly kept the reins. Unfortunately, no single adviser stood out. Without a clear methodology, Ayden trawled through her file.

W.R.R. MUNRO

14:
JUNE 2034

A farmer stood looking over a barren landscape. His khaki trousers and blue work shirt were covered in dust. He removed his hat and looked directly at the camera.

"We used to grow green beans, cabbages, cauliflower, pumpkins, oranges and peaches." He glanced around again. "Far as you could see in every direction."

He kicked at the white crust at his feet.

"Not any more. Our parents irrigated too much. Replaced too many trees with shallow rooted crops. It's only good for modified salt tolerant wheat now." He wiped his brow.

"Which is just great...expensive seed and low yields."

NAROOMA NEWS ONLINE, "DAMAGED BY SOIL SALINITY AND ECLIPSED BY WESTERN AUSTRALIA'S NORTH, THE MURRAY DARLING BASIN CAN NO LONGER CLAIM TO BE AUSTRALIA'S FOOD BASKET," NEWS.NAROOMA

Thursday, June 1st

After nearly forty-eight hours, Ayden found himself sitting in his over-designed office, staring at a photo of Sandra Allen standing beside some very happy people holding up one of those huge media checks they create for publicity. It was probably a meter and a half wide. Ayden had hardly slept and been surviving on canteen sandwiches delivered to his desk for two days...and he had nothing to show for it.

He sighed. She was exactly what she looked like, a very wealthy, very intelligent philanthropist. Ayden stood and stretched. He would have to admit defeat. The problem was, he wasn't at all sure what to do next, and the deputy director was starting to get impatient.

He looked at the clock floating in his scape background. It was three in the afternoon. He'd worked until eleven the previous night and been back at it from seven in the morning. He needed a break. He needed a dose of irreverent attitude. For the hundredth time, he resisted the urge to call Tess.

Flick. [James?]

[Yes, sir.]

[Feel like some fresh air?]

[Always, sir.]

[I need to clear the cobwebs. I'm thinking of getting out of the city completely for a couple of hours and I understand there are good walking tracks around. Think you could rummage up a car for us?]

[Let me come back to you in just a few moments, sir. I'm sure we can organize something.]

Ayden could, of course, have organized a vehicle in about the same time it took to ask James to do so. What he was really doing, was asking for security clearance.

[Mr. Walker, sir? A car and driver will be downstairs in fifteen minutes.]

[Thanks, James. Now I hope you have a change of clothes handy. I'm not sure that Italian suit of yours is the best choice for –]

Ayden broke off, staring into his specs.

[Sir, are you okay?]

Ayden heard the sudden physical exertion in James's voice and was about to say something when the door flew open.

"Sorry, James, I'm fine, but please cancel the car. Sorry to disturb you."

Flicking the photo to a massive scale with two fingers, he examined it closely, no longer conscious of James, who hovered uncertainly in the doorway then quietly left. The check wasn't in her name, of course. It was from the Bill and Sandra Allen Foundation.

The signature wasn't hers, either. Like all signatures, it was hard to read but it definitely didn't say Sandra Allen or S. Allen. The first initial was definitely an "N" and the surname started with a "T." It was, of course, just a publicity prop and hardly anyone actually used handwriting signatures to authorize payments any more so it was possible the name was fictional as well but...

Henri was incommunicado, as was frequently the case. She never offered to tell Ayden what she did when she dropped out of sight and he never asked.

[I need to talk to the deputy director asap,] Ayden messaged to the deputy director's secretary. His phone chimed within seconds.

[Hello, Ayden. Please clarify. I have instructions to interrupt the deputy director if you say it is very urgent but believe me it will be better for us both if I don't do that unless it's a real emergency. Is it an emergency?]

"Oh. Um, no." Ayden smiled. "Sorry Wil." Wilhelmina—Ayden was never offered a surname—was always charming. To speak to her, you would never guess who she worked for or that she had any pressure in her life. "It's not that urgent, but please ask him for two minutes today some time, if possible."

"Will do. Can I tell him what it's about?"

"I want to ask for access to bank records for a charitable foundation."

"If that's all you need, would it be okay for me to get back to you?"

"Yes, Wil, fine. It's the Bill and Sandra Allen Foundation I'm interested in."

Wilhelmina called Ayden back ten minutes later. "I'm authorized to give you access, Ayden. I'm sending you a link for financial and legal records for the Foundation going back to its inception and the same for Sandra Allen's investment vehicles. The link also provides you with access to lots of other information about these entities."

"That was quick. Thanks, Wil."

"My pleasure, Ayden. The deputy director also wanted me to note that as Sandra Allen is very high profile and extraordinarily well connected, we need to tread very softly. He instructs you not to discuss this with anyone other than him, me or Colonel Soulandt."

"Understood. I haven't and won't."

"And the deputy director wants you in his office tomorrow morning to present a progress report. The only slot I have available is 6:15 a.m. Sorry."

That gave Ayden fourteen hours if he needed it, assuming he allowed himself four hours to sleep. He had access to everything—bank account statements, authorized signatory lists, scans of physical documents, trust deeds, and board minutes.

There were seven authorized signatories for the Foundation account. Six of those were authorized only as joint signatories. Only someone named Nick Thompson could authorize payments alone and he had authorized over fifty percent of the Foundation's key research grants. Ayden called up records for Sandra's main investment vehicles—the Sandra Allen Trust and SBAC Investments. Nick Thompson was again a sole signatory and authorized many of their larger investments. Ayden dived through the bank records.

Retinal scans were the norm for high value electronic payments. They were terrifically difficult to fool so the security they offered was hard to match. Thompson's biometric of record was his fingerprint. This wouldn't be unusual in a man of, say, sixty, who had never bothered to update his biometrics and who rarely made payments of more than a few thousand dollars, but Thompson was ostensibly thirty-two. His PIN was eighteen characters long, which would have offered some comfort, but Ayden couldn't imagine the Foundation Trustees would have been happy with their only sole signatory using fingerprints.

Ayden felt energy course through his arteries. On the face of it, this lead was no better than many others but it somehow felt right. Thompson was an obscure lawyer, very much behind the scenes but wielding inordinate control.

Equally interestingly, biographical information was sparse. Though a lawyer, Thompson's role was essentially a top-level investment gatekeeper for Mrs. Allen's investments and for the Foundation, but he never attended any of the interminable industry functions or launches. There were no photos in the public record. He had never married. He had no children. He wasn't listed as an active member of any clubs or organizations. The only photograph Ayden was able to access was Thompson's passport photo.

Ayden peered at the tiny photo. Thompson's hair was light brown, long and foppish. He looked younger than his claimed thirty-two years. He wore a conservative suit, a bold striped shirt and a bright floral design tie. The overall impression was of a fashionable young man, a marked contrast to Hanford's more severe persona. Even aside from the soft chubbiness, his face wasn't quite right. Could be, he was just a workaholic who valued his privacy.

Nevertheless, it was Hanford. Ayden was certain of it. He spent six hours assembling his case, had a quiet evening meal by himself, then went to bed early.

Friday, June 2nd

It was hard to tell if the deputy director was actually pleased. He looked as if he had been awake for hours, but he was sipping coffee. The big wall screens at the far end of his office showed various photos of William Hanford and the passport photo of Nick Thompson.

"A team of thirty plus experienced intelligence analysts couldn't find him, but you have…essentially on your own. That's what you're telling me?"

"I couldn't have done it without the team."

"Your efforts to share the glory are noted, Walker," the deputy director said. "More to the point, are you sure this is him? Doesn't look quite right to me, and the face recognition software only gives it a 37 percent likelihood." The deputy director held up his hand. "I know, I know, the photo could have been manipulated so it's close enough for a cursory glance but designed to confuse the face recognition software. That works as long as he doesn't want to travel on it. I can't believe we don't have a single high-res shot. Irritating. What I want to know is, what makes you so sure?"

"It's him. That's why you don't have high-resolution photos. What convinces me is his profile. It's perfect, very much behind the scenes but pivotal. Easy to maintain while he was Hanford. What worries me is that he may have another one of these. I think we have to move quickly."

"Obviously. But Hanford identity or not, he's important to Sandra Allen. Anything happens to him, she'll raise an almighty fuss. We don't move 'til we get a good, recent photo. And until that happens, you talk to no one about this."

"Understood."

"I will have a very small, select team surveil this Thompson character and provide us with photos. I'll have a crash team ready but they won't know their target 'till they get the go. Be ready to review the photos and to move on a moment's notice."

"Okay. Move?"

"It's going to happen very fast and, though I can't think what for, I want you there. He isn't going to slip out of our hands again."

Sunday, June 4th

Ayden was eating a light salad at his desk when the video came through some thirty hours later, the delay a testament to the low profile Thompson was keeping.

It showed him getting into a long, black limousine. Ayden was astonished. He almost doubted his own identification. Could Hanford have managed to become obese in such a short time? Why would he? Nine weeks previously, Ayden would have estimated Hanford's weight at a whip-thin one hundred and sixty pounds. The man in these photos had to be well over two hundred pounds. That sort of weight gain would have required concerted effort and presented a significant health risk if maintained. Would Hanford really expose himself to the health risks for the sake of disguise?

The fat layer also caused difficulties for the facial recognition software, which only calculated a fifty three percent likelihood of a match. The deputy director had a forensic artist start with a photo of Hanford and morph a major weight gain. Now, the software estimated the match at over 80 percent.

• • •

Four hours later, Ayden was in Texas.

He was bustled into a fast sedan, then a helicopter, then into a hulking four wheel drive. Thompson lived in a two-story penthouse at the top of the Harkness Building in downtown Houston.

As he traveled, Ayden followed the crash team's progress. Thompson hadn't been sighted again since the video was taken. The limousine had driven straight to his apartment building's underground car park, where it had disappeared into his private parking area. An agent had been placed into the building security control center where he watched as the building's cameras captured Thompson in his private elevator, being whisked to the top floor. The car had been intercepted and searched as it left the building, the driver detained. Agents were stationed at ground floor elevator doors, fire escapes and the garage door of Thompson's private parking area.

• • •

Ayden was shown to the doors of a large van parked outside the building, inside which he found Henri waiting, wearing a fierce smile.

"Hi, Ayden, long time no see." Inside, the van looked quite bare. Padded bench seats ran down either side, underneath which equipment lockers were built in. Opposite Henri sat a small-framed man with very short, ginger hair and an intense expression on his heavily freckled face, completely absorbed by his specs. Behind them, towards the front of the van, were two more men, both wearing fully immersive VR helmets, fingers twitching but otherwise immobile.

"This is 'Jones.' Team leader for the grab." Henri indicated the man opposite her. She slid down a little, patting the seat beside her. "We've saved a seat for you. Make yourself comfortable." She ignored the two helmeted figures in the background, as though they were part of the equipment.

As soon as Ayden sat down, someone outside closed the doors. The seats were unexpectedly comfortable. Good lumbar support. No doubt agents often had to spend a lot of time on them.

"Hi, Henri. Hello, Jones, pleased to meet you."

The team leader, his specs remaining opaque, gave Ayden a distracted nod and flicked him an icon. "You'll need this." Without turning away, Jones's attention clearly departed. "Right. Look alive people. The eyes have arrived so this is it. Go, go, go," he commanded quietly.

Ayden flicked the icon open and watched from various vantage points as agents simultaneously entered the stairwells from every floor. Six agents rode up in Thompson's private elevator and used security keys to enter the apartment.

● ● ●

The deputy director was furious.

"How is it even remotely possible he isn't there?" His huge avatar glared down at the team leader. "I want that entire building searched, every damn inch. Every apartment, ventilation shafts, crawl spaces, underwear drawers, everything."

"Yes, sir, but…"

"Don't you dare not find him Jones. Soulandt, get a hundred FBI agents there pronto and assign them to Jones. Only ONI people inside Thompson's apartment though. Declare it an anti-terrorist swoop. Call in local police and SWAT to make it look real but keep them outside the building. Hell, let off a flash grenade and smoke the place up if you need to, but find him."

"Yes—"

"I'll be there in a few hours. Jones, make me happy! Walker, assume that Jones there will continue to screw it up. Do whatever magic you do. Find Hanford. Soulandt, give Walker whatever he needs." The deputy director disappeared from their shared scape.

No one spoke or moved for several moments.

"Great," Jones said under his breath. He sat back and stretched his neck, took a deep breath, then leaned forward and started firing orders at his team. "White! In here now. Blue! I want two perimeters..."

Henri raised an eyebrow at Ayden then began making calls.

Ayden watched the frenzied activity on the bank of windows arrayed in his specs.

● ● ●

Ayden and Henri relocated to a somewhat run-down hotel a block away, leaving Jones to his van. The Office of National Intelligence had booked a suite of rooms, much to the delight of the hotel management.

The largest suite, the honeymoon suite, had been invaded by an army of assorted, oddly shaped suitcases which plugged into power outlets and filled the room with the gentle whisper of cooling fans.

Ayden found a slightly shabby but comfortable chair and spent a few seconds gazing at the spectacular sunset, courtesy of one of Houston's regular summer dust storms, then he opaqued his specs and watched the ONI team at work. They took Thompson's apartment to pieces. They searched every one of the eighty-seven apartments in the building. Lawsuits would follow. They searched parking lots, storage rooms and maintenance spaces.

Ayden switched back to the surveillance footage. He watched Thompson emerge from the offices of Crawford, Hoskins, and Belding, a prominent Houston law firm. He watched Thompson climb into the limousine. He watched an unbroken sequence of footage of the limousine driving to the apartment building, entering the underground parking and disappearing behind the garage door to Thompson's private parking area. He watched building security video of the inside of the private elevator as Thompson entered from the parking lot, pressed the "Penthouse" button and stood back as the doors closed and the elevator began to move. He watched Thompson wait, then depart the elevator as the doors opened onto his private foyer. Simultaneously, he reviewed the elevator data, showing it ascending through the building and stopping at the top floor where it remained until the ONI team called it down.

The deputy director swept into the room with Jones trailing uncomfortably behind him.

"So what did he do, hang-glide off the roof?" the deputy director said. "Rappel down the building exterior?"

"We don't know, sir," Jones said almost inaudibly.

"Goddamn it, man, you wouldn't be here if you were incompetent, but this should have been a walk in the park."

"Yes, sir."

"I very much hope, for your sake, that he did something exceedingly clever."

"Yes, sir."

"Walker! You're supposed to be my trump card. How did he get out of that apartment?"

"I don't know but we underestimated him," Ayden said calmly. "He *has* done something exceedingly clever and unexpected. He's implemented an elaborate contingency plan. What we need to do now, is be careful not to make assumptions. Perhaps, he never entered the apartment."

"In other words, you haven't a clue. We have footage of him entering the apartment, Walker."

"It certainly seems that way, sir. I'm sure his obesity is somehow a clue but I need to think more about it."

"He got himself fat to evade our face recognition software, Walker. You're grasping at straws. Go away, all of you. Find him."

• • •

Ayden retraced Thompson's movements, had an agent commandeer a car and take him to Thompson's office, then back to the apartment building and into the parking area, simultaneously running the surveillance footage. He rode the private elevator up to the penthouse. He wandered through the apartment, now a scene of complete chaos as agents conducted their third, and even more destructive, search.

It was perfect for a recluse, had either been built that way or heavily modified. That wasn't surprising. The very wealthy and celebrities alike had to take extraordinary measures to give themselves even a modicum of privacy. As he wandered, Ayden looked into the building's history. Translucent icons and documents hovered in front of the bustle. The building was only ten years old, an Abacus Inc. development. Like all property developments, it was funded largely through debt.

Ayden dug into the financing. Standard bank lenders. The build had been very expensive, however. It had exceeded the conservative bank valuations and they'd required additional security, which came in the form of an additional mortgage over another property. Ayden traced its ownership to SBAC Investments, one of Sandra Allen's investment trusts. He froze, considering.

"You look like a rock in the middle of white water rapids," Henri said, walking into the apartment.

"Henri. Hi. Listen, assume Hanford was behind the building's construction. That presents possibilities, doesn't it? Where's Jones?"

Ayden and Henri got together with the task force leader in the building's security control room. Jones had made it his de-facto base of operations. Ayden suspected this was his way of trying to steer clear of the deputy director.

"There's just no way he could have gotten out of that apartment," Jones said. "I've been doing stuff like this for nearly twenty years, and there's nothing difficult about containing a building. I had men in the stairwell, men in the lobby watching the elevator's only other exit point aside from the apartment, and men in the garage, just outside the entrance to his private parking area. I even had surveillance drones flying around the top of the building."

"Finished your measurements?" Henri said.

"Yes, Colonel. There isn't a centimeter in either of the two top floors unaccounted for. No bolt-hole, no escape shaft."

"Did you have men inside the elevator well?" Ayden said.

"No, but the video shows him exiting the elevator in his private lobby, then the elevator car stays empty and, well you've seen his place, we've torn it apart. And there's no other access to the elevator shaft from this floor."

"Okay, let's have another look at the tapes," Ayden said.

A small sigh escaped Jones.

Henri's response was immediate. "You got something better to do with your time, Jones?"

Jones locked eyes with Henri. "No, sir…"

Henri waited.

"But, I've watched that footage a hundred times. I know it better than I know my own face. Sorry, sir." Jones reached a finger to the icon.

"Wait," Ayden said. "Let's look at the footage from the previous time he entered the building."

Jones and Henri just looked at him.

"You can tell us if there's anything different."

"Okay, well, according to the log that was the previous evening."

They watched the video. "Nothing," Jones said. "It's the same."

"Looks that way," Ayden said. "Mostly. Let's run the two simultaneously, side by side on two windows."

They watched again.

'There! See where he exits the elevator? He seems to veer just a little left, rather than right, before we lose sight of him? Again, from the start."

They watched again.

"And he swipes his security card from top to bottom, see? The previous evening he swipes it from bottom to top."

Henri looked at Ayden. "Could the sensor be sophisticated enough to read the difference?"

Jones answered for Ayden. "No reason why it couldn't have been built that way."

As one, they rushed for the elevator. Ayden swiped the card down and pressed the "Penthouse" button. They waited breathlessly as the elevator ascended…and opened to Thompson's foyer.

"Uh huh," Jones said.

Ayden walked out of the elevator, veering slightly left. "It's wrong. Veering left takes him nowhere," Ayden said, stopping at a blank wall.

"It's solid," Jones said, tapping at the wall.

"There's something else. We're still missing something," Ayden said.

"Wait!" Jones's body was rigid, his eyes staring into his specs. "There is one other thing. It's tiny. It can't be anything, but…"

"But no sillier than swiping the card the other way?" Ayden said.

Jones smiled and they rushed for the elevator again. Ayden handed him the card. It was the first time he'd seen Jones smile.

"Just a moment," Henri said. "What exactly are we expecting might happen here?"

Ayden smiled. "No idea."

Henri drew a pistol from inside her jacket. "Okay, Jones." She nodded to him.

Jones swiped the card downwards, talking as he did so, and looking at something in his specs. "As his index finger presses the 'Penthouse' button, I think his pinkie finger presses the 'Parking' button at the same time." The elevator started to descend. "You see it when you run the footage simultaneously."

No one spoke as the elevator descended. As they felt it stopping, Henri stood forward with her pistol at the ready. The doors opened. Nobody moved.

Ayden realized that his mouth had actually dropped open. He looked at his companions and laughed. "We're not in Kansas anymore."

"So where the hell are we?" Jones said.

Still, nobody had moved. The doors began to close. The sudden movement snapped them out of their inaction. Henri slapped the door open button. Ayden quickly stepped out of the elevator. Jones barked into his communicator. "Deloitte, are you in the security room? Where does the system show the elevator is right now?"

Ayden was examining the walls of the room he had entered.

"Whoa," he heard Jones say.

Ayden looked back at the agent.

"He says we're in the parking garage."

"Jones. Back in that elevator," Henri said. "Swipe the card up. Press the 'Parking' button. Assuming that does take you back to the parking level, do your down swipe and double press thing again. Move!"

Ayden looked around again. They were in an exact copy of Thompson's private elevator foyer, except that the polished wooden door was narrower than the one which led into Thompson's apartment and was positioned to the left of where it should have been. From the limited view of the security camera mounted inside the elevator, it would look identical.

Ayden reached for the door handle.

"Wait," Henri commanded.

The elevator door opened to a smiling Jones. "According to the elevator system, I'm now at the Penthouse level," he said, his smile widening further. "And the elevator descended very slowly, took as long as it takes to get to the top."

They all looked at the wooden door.

"I'll get a crash team down here," Jones said.

"Why?" Ayden said.

"The door, or whatever is beyond, could be booby-trapped," Henri said.

"No," Ayden said. "Not Hanford's style." He stepped forward quickly and opened the door, as Jones hauled him backward and Henri rushed forward, thrusting her pistol in front of herself.

Nothing happened. "How do you want to play this, Colonel?" Jones said quietly.

"His instincts have been perfect so far," Henri said without turning around. She kept her pistol pointed down the empty corridor. "And the DD doesn't seem in the mood for delays."

Jones sighed. "Well, if you are determined for us to do this like amateurs, at least let the expendable guy play cowboy." He pushed gently past Ayden, stepped around Henri, and examined what lay beyond.

The corridor was long, low and narrow. A strip of LED lighting ran along the center of the ceiling. The walls were featureless, off-white panels. At the far end was another door. He spoke without turning his head. "Please wait here until I have traversed the entire corridor. If it is booby-trapped then whatever happens may not happen 'till I open the far door. Will you do that for me, Ayden? Please?"

"Yes."

Jones inched down the corridor, with every step peering at the floor, walls and ceiling. After what seemed like an interminable wait, Jones reached the far door. He paused, looked back at Henri and Ayden, flattened himself against a wall, reached for the handle, and slowly, silently opened the door.

Nothing happened. Jones peered beyond the door and gestured them forward. It was obviously a larger room, unlit. A few feet in front of the door was a wall of cardboard boxes. Ayden pushed experimentally on the stack in front of him. Whatever was inside was heavy. He tried the next stack, the one against the wall. It moved easily. Carefully, shoving mainly on the bottom box, he pushed the stack out of the way. They were in a small, very standard storeroom. Ayden tried the handle of the door opposite. It spun freely, the door remaining firmly locked. Reaching into his pocket, he swiped the security card past the door, near the handle. It clicked open.

From the outside, it was a blank metal door, like any of the other utility doors in the corridor. Jones remained inside the storeroom. Ayden and Henri walked past restrooms into a crowded shopping mall.

It was Henri's turn to mutter, "Great."

The average noon temperature of Houston in summer was eighty degrees Fahrenheit. In direct sunlight, effective temperature regularly exceeded one hundred degrees. The business people in their suits and the shoppers in their "see and be seen" outfits objected to being exposed to such temperatures and the resulting flood of perspiration, so the city of Houston adapted. Nearly every building in the central business and shopping district was connected to a network of over twenty miles of air-conditioned underground walkways.

There was, of course, extensive security camera coverage of the mall and connected corridors but none aimed specifically at the point where the utility corridor entered the mall.

The team directed to review the footage from nearby cameras found nothing.

Hanford was gone, again.

Monday, June 5th

The deputy director was calm.

"You were right, Walker. It wasn't incompetence. That would have been unforgiveable. Hanford out planned and outfoxed us. I underestimated him. There's nothing on any of the building or mall plans to indicate the existence of the extra elevator level or the corridor to the storeroom. That level of planning and preparation is worthy of a government agency, rather than of an individual. From now on, we treat him like a well funded state intelligence organization. That means he could be anywhere on the planet. He could hole up in a cave for years. It took us a decade to find Bin Laden. It may take us as long to find Hanford."

The deputy director lapsed into silence, staring out the penthouse windows. The electronics-filled suitcases hummed quietly to themselves and the room smelled faintly of ozone.

Jones, his credibility somewhat restored, had joined Henri and Ayden in the meeting. "If we're in for the long haul, sir," he said. "Do we consider bringing in outside agencies?"

"Not yet, Jones. I need to give careful thought to what we tell them." He eyed Ayden hopefully. "Do you have any sparks of brilliance you can throw onto the table, Walker?"

"There's still a piece that doesn't fit. I don't believe he got himself so obese just to evade the face recognition software. There are other ways and the health risks are significant."

"Yes, but only if he maintains the weight for a long time," Henri said.

"It doesn't explain the weight sufficiently," Ayden said. "Not to me."

Henri fell silent.

The deputy director peered at Ayden. "Okay, Walker, I'm not going to write off your instincts this time. Soulandt, get a team together. Analysts, scenario specialists, grab whoever you need. Alpha minus priority, on my authority. Support Walker anyway he needs. Look at all the off-the-wall options. Daily reports."

Friday, June 9th

Ayden peered into his specs. "Wow. This is different. Clearly, you don't suffer from vertigo. Where's it from?"

"Good different or just different?" Henri said.

"It's beautiful. How do you get access through mil-scape and whose work is it anyway?"

"Ah, it's amazing what you can do in mil-scape with admin rights but this isn't simulated. It's real. Real-time, even."

"Amazing filters, then. I didn't know this was possible. Real time? Must use scary amounts of processing power."

"Think military. Quadruple redundancy and too paranoid to offer the massive amounts of excess capacity outside mil-net. Actually, IT encourages us to be as demanding as possible. It'd all disappear in a picosecond, of course, if the proverbial hits the fan, but I've never seen it happen."

Ayden looked around. "So, you're filtering out the water?"

"Yes."

"But not the ripples on the surface."

"You going to analyze it to death or just enjoy it?"

Ayden smiled, tried to look everywhere at once. The wall was about twenty degrees from vertical and it didn't quite reach the surface.

And it was alive. He could hardly see any rock. Hard corals, soft corals and every sort of plant vegetation imaginable competed exuberantly for space, and fish in their thousands hid in crevices or floated impossibly past the multicolored surface. Far, far below them, the wall terminated in a jumble of what must be building-sized rocks. Away from the wall, the sandy sea floor stretched forever.

Looking away from the wall, Ayden saw turtles and huge manta rays and sharks, and out there, in the mist of distance, larger shapes moved. He saw a whale and even larger but less distinct, a huge shoal of fish.

"I've done some scuba diving but it was nothing like this. You can't see very far at all."

"I dived a lot as a kid," Henri said. "I was a navy brat. One day, off Hawaii, a particularly clear ocean current came through. I nearly died of cold but it was worth it. They said it was eighty meter plus visibility. It was enough that you felt more like you were floating high in the air rather than underwater, but even that was nothing like this."

"So, the military is paying for a series of high-resolution stereo-scopic underwater cameras to sit just off a coral reef and look at it? For your entertainment?"

Henri laughed. "Hardly. The cameras are part of a tourist in-stallation but the tourists get the usual watery view. Mil-scape takes the raw feed and re-creates the view without the obscur-ing water. That reminds me, this is classified, it's well within your clearance level, but you're not allowed to tell anyone."

"The magic's in the light." The programmers hadn't filtered out the light playing on the surface. Millions of ripples and wavelets interacted chaotically above their heads and the sunlight turned it into an infinitely varied, endlessly changing ceiling of light and shadow, an interplay which was also reflected on the living wall beside them.

"Has a military application, of course," Henri said. "It's very handy for our subs to be able to see clearly, especially in the dark depths. For all I know, I'm the only person who's applied these filters to shallow water feeds."

"It's wondrous. Now I can see how you survive long stretches in the cave. It's funny. I didn't take you for the type to slow down and smell the flowers."

"I find it sort of meditative. I soak it in, clear my conscious mind. Let my sub-conscious do the heavy lifting for a while. Im-proves my productivity. Well, that's my story anyway and I'm sticking to it."

"Fair enough."

"I can give you access, if you'd like?"

"Oh, yes, please."

"Speaking of productivity…"

"Ah, yes. I'm not really ready to report on this yet but I'm inves-tigating chemically induced metabolic depression."

She peered at him for a few moments, then burst into laughter. "Oh, the DD's gonna love this one. You're talking about hiberna-tion? You're suggesting Hanford's hibernating like a bear in his den?"

Ayden sat back and tried to look offended but a smile creased his eyes and in any case, a fish just swam through his head. "Actually, some biologists argue that bears aren't true hibernators. Although their heart rates drop dramatically and they can go for three months without food, water or exercise, their internal body temperature doesn't actually drop very far, compared with, say, a hibernating arctic ground squirrel whose internal body temperature can drop below freezing."

It just made her laugh harder. She started to curl up, holding her belly.

Clearly struggling, she controlled herself enough to speak. "Just do one thing for me, please, please let me be there when you pitch it to the DD," she said and collapsed into laughter again.

Ayden's smile widened. He waited, content to watch her mirth. Eventually, she calmed down and wiped the tears from her eyes.

"Thank you, Ayden. That was the best laugh I've had in ages. Wonderfully therapeutic."

"Just so you know, I am serious."

"Oh, I don't doubt it and you have more chance of selling the idea to the DD than anyone else but still, if human hibernation really had legs, don't you think NASA or DARPA would be looking at it? They're throwing serious money at popsicle research."

Although a "permanent" moon base had finally been established, none of the proposed manned Mars missions had launched. As experiences aboard first SkyLab then the ISS had revealed, humans were poorly suited to the psychological and physiological rigors of extended stays in isolated, irradiated, microgravity environments. The life support infrastructure required to keep them healthy for the two year round trip to Mars was unjustifiably expensive in these economically challenged times. Tuck the astronauts into tiny coffins, however, and you drastically reduce the life support infrastructure requirement… and the worst psychological issues are avoided. Storing them behind a water tank would provide protection from irradiation, and frozen—or hibernating with radically reduced metabolisms—their bodies wouldn't adapt to the microgravity environment. As a result, NASA spent hundreds of millions of dollars per annum funding cryogenic suspension research. DARPA joined in, seeking the benefits of freezing battlefield casualties for transport to fully equipped medical facilities. Even the private sector was involved, offering rich, desperate people with terminal illnesses the hope that medical science would advance sufficiently to both revive them safely and cure their illness.

Nevertheless—and despite the existence of a 14 trillion dollar per annum international biotechnology industry—there didn't seem to be much interest in funding hibernation research.

"Actually, that's sort of my point. Hibernation looks a lot less dangerous than freezing and the attractions to NASA and DARPA should be the same but they aren't looking at it."

"The most likely reason being they've tried it and it doesn't work."

"Nor does cryogenic suspension so far. Hasn't stopped them from trying. Let me run this past you. Firstly, Hanford has a track record of suppressing research he's particularly interested in. He likes to make his advances in private. Second, there are lots of close human analogues. Whereas only frogs are known to freeze completely then thaw and wake up, many mammals hibernate and the basic processes are reasonably well known. Even better, humans and bears have similar physiologies and the relatively high temperature version of hibernation that bears use is less risky than the low temperature hibernation of the smaller mammals and it provides a much faster recovery time. Third, it's an excellent fit with Hanford's sudden obesity. Four, the infrastructure requirements are low. The necessary drugs and delivery systems could probably be stored in a suitcase with minimal energy requirement as long as a supply of fresh air was assured. He could have set up multiple bolt-holes, each equipped to offer a place to disappear in an emergency. He could then simply 'sleep' for a month or two, time for the initial search intensity to pall. Five, risk management. Where a power failure would quickly kill a popsicle, if there was

somehow a failure in the cocktail of drugs likely required to maintain hibernation, he'd presumably just wake up. So, thoughts?"

"Hmm," said Henri then she stared in the direction of the wall
for a while, though seemingly oblivious to its vibrant life. "You
know, Ayden, for a scientist, you're pretty happy to pile wild ideas
on top of speculation without having a shred of evidence to point
at."

"I'll let you in on a little secret, Henri. All scientists do it, even
your military ones, though they may not admit it. That's the fun
part. The work part is digging up the evidence."

Ayden had quite enjoyed reading, what research was available,
regarding hibernation. No one had admitted experimenting with
humans or any other primate, of course, but there were studies
of the biomechanical and pharmaceutical process in animals and
he'd pieced together the chemicals likely to be needed to study
and test hibernation approaches. What remained was to track
down the suppliers of biomedical products worldwide and follow
up each and every purchase.

Fortunately for Ayden, there was a seemingly endless supply of
analysts who could be assigned to the mind-numbingly tedious
task.

Sunday, June 11th

It could never have happened with civilian gear. One moment,
Ayden was deep in thought, fine-tuning his analysis, shuffling
associational icons. The next, with no warning, no transition to
soften the shock, he was sitting in a pixel-perfect simulation of the
deputy director's office.

It was an extraordinarily disconcerting experience.

"I have a five minute window. What have you got for me?" the
deputy director said.

Ayden was trying to collect his wits and frame an answer when
he realized the DD wasn't looking at him. One of the drawbacks
of simple specs, as opposed to a full VR enabled room, or a VR

helmet for that matter, was the lack of peripheral vision. He was turning his head when Henri spoke. She appeared in his specs, as if from nowhere, right beside him. Ayden jumped.

"Yes, sir," she said. "We quickly narrowed our analysis to two sets of possibilities. The first is cryogenic suspension, but unless Hanford has managed to achieve something which nearly half a billion dollars of NASA and DARPA money haven't been able to match, cryogenic suspension remains essentially a drawn out form of suicide. It is possible that—"

"Skip it, Soulandt. I do glance at your dailies and I don't have time for the long version. You're looking at hibernation, something even the cowboys over at DARPA haven't seen fit to fund. Explain."

Henri had to know the deputy director wouldn't listen to a preamble. Either she was nervous about presenting Ayden's theory or she also needed a moment to adjust to finding herself suddenly sitting in front of the deputy director. She glanced at Ayden, fractionally raised her eyebrows.

He nodded minutely.

"Yes, sir. We think the hibernation idea has legs. With your permission, sir, I'll let Ayden explain."

The deputy director turned his gaze to Ayden.

"I'll assume you've either read or don't want the details," Ayden said. "But there are traces of evidence to suggest that, for nearly a decade, Hanford may have been burying research with possible applications to human hibernation, even leap-frogging DARPA just by offering funding with less hoops to jump through. If he's followed his father's methods, he'll have signed up his researchers to confidentiality agreements then kept each research strand separate so no one other than him knows the full picture."

"I'm prepared to assume he's capable of it," the deputy director said. "What else do you have and how do we find him if you're right?"

"We follow the molecules. Animals produce delta opioids which depress their metabolisms at a cellular level. We've come up with a list of chemicals Hanford might need and one substance presents itself as an excellent candidate. The delta opioid DADLE could well be vital to the process but it doesn't like freezing and, even stored carefully, deteriorates over twelve to eighteen months. So, if Hanford has been maintaining multiple bolt-holes, he'll have needed an ongoing supply."

"Sounds like a bit of a stretch."

"We tracked a few other molecules as well and got nowhere but DADLE doesn't have very many applications and our team has uncovered ongoing orders from several labs within the U.S. where the need isn't obvious. Two of the labs are associated with Genenco funded research."

"That sounds a little stronger. What are we doing to confirm it?"

It was Henri's turn. "We've infiltrated the labs. So far, can't find any positive evidence, but the hormone isn't there and they have no legitimate use for it, so why the orders?"

"Okay, Colonel, so what are we doing to find bolt-holes?"

"We're just starting to plan our approach, sir. We'll start with the labs, of course. Try to track where the hormone went. Examine building plans for the labs. Make our own measurements, looking for gaps. I'm not optimistic though. The key advantage of the hibernation technology, assuming that's what he's doing, is that it could be set up anywhere. He could be in any private house or apartment anywhere on the globe by now."

"Hmmm. Walker, thoughts?"

"I haven't had a chance to chat through this with Henri yet but I think the whole point of Hanford setting up hibernation bolt-holes is so he could disappear, immediately and without risking capture by travelling, so I think we have to look closer to home."

"Specifically?"

"Well, your staff has watched every second of the security footage from every camera in the mall. It really doesn't look like he left the utility corridor."

"You mean the one his secret passage opens onto?"

"Yes, sir. I think we need to retrace his steps from the elevator. Physically explore for a bolt-hole from the elevator door to the mall corridor to start with. We could, of course, just start ripping down the walls but I don't suppose we have some sort of sonar we can use to detect cavities?" Ayden said, looking from the deputy director to Henri.

"Good question," the deputy director said.

Monday, June 12th

They had a cavity sensor. The CIA developed it when they were searching for Saddam Hussein back at the turn of the century at the end of the second Gulf War. It was used again in '22 when the FBI famously caught the Great Lakes Killer just as he was about to kill his nineteenth victim in a hidden room of his house.

The scanner consisted of a panel that had to be held against the relevant surface. The results were sent to the operator's specs, which displayed both the raw data and a graphical representation of wall density. In practice, it meant two operators taking turns to slowly wipe the heavy panel across every inch of wall, ceiling and floor while four response team personnel stood around waiting for something to happen.

It was excruciatingly slow. They took over three hours to scan the elevator antechamber. The corridor took another three hours. By then, the operators' arms were shaking with fatigue. The bored response team was co-opted to assist.

Scanning the storeroom at the end of the secret corridor they got an almost immediate result. In the space behind where the stack of boxes had been, a section of wall showed a cavity the full height of the room and two feet wide. They called in the result and waited.

• • •

Henri and Ayden raced to the building. When they arrived at the storeroom, they found the response team standing around a section of wall, fully kitted out with flack-jackets, helmets and various weapons.

"We have strict instructions, sir," one of the men said to Henri. "We go in first and clear the area."

"Don't worry, agent. We'll do it by the book. But first, let's see if we can open the door quietly before pulling the wall apart."

The entire room was lined with cheap pre-fabricated panels. Ayden examined the wall closely. It looked like some sort of bio-board, the ubiquitous compressed cellulose panels which had replaced fiber cement boards when their carbon sequestration value became clear. The producers earned carbon credits growing the various plants used to manufacture the boards, and they weren't penalized when they harvested the plants, because the carbon would be locked away in the boards and the plants replaced by the next crop.

The gaps at the edges of the panels were too narrow to provide purchase for fingertips, and there was no sign of the panel having been pried. Ayden stood back, bumping into the response team agent who had been peering over his shoulder. He considered the situation. Hanford would have planned for access in an emergency scenario. Ayden reached into his pocket for Hanford's security card and wiped it past the right hand side of the panel where the handle would have been if it were a door.

The panel didn't budge but Ayden thought he heard a faint click. He wiped the card past the door from right to left, above head height. This time he was sure he heard the click. Then down the left hand side, and the panel clicked open just a crack.

Gently but quickly, the response team leader inserted himself between Ayden and the panel opening.

"Sorry, sir, please step back," he whispered to Ayden without taking his eyes off the panel.

"Come on, Ayden," Henri whispered, smiling. She was standing at the back of the small room. "Let them do their jobs."

Ayden watched as one agent whipped the door open while a second, crouching low, thrust flashlight and weapon into the dark cavity beyond. It looked like yet another corridor, albeit only two feet wide. It was short though, terminating in a blank wall. No, the corridor turned left at the far end. The response team leader entered carefully, sideways, inching toward the corner. As he disappeared around the corner, a second agent rushed in to stop at the corner, keeping his leader in sight.

"Say again," said the agent beside Ayden, pushing his earpiece more firmly into his ear. He looked at Ayden. "Another door, sir, locked. Can I have that swipe card please?"

Ayden found himself holding his breath. He let it out with a smile.

"Exciting, isn't it," whispered Henri. "Could you see yourself going back to a normal life after this, assuming you could?"

Ayden was pondering his answer when the final agent in the room with them spoke. "It's clear for you to go through, sirs."

Ayden and Henri rushed to the opening.

"After you, Henri," Ayden said.

"No, Ayden, you've earned first peek, now that it's safe."

Ayden rushed down the corridor, rounded the corner, down the next corridor, then through the open door at the far end. He emerged into a large elevator well. It served six elevators. Behind him, Henri groaned.

"Oh, gawd. He could have gone anywhere from here. It'll take us a week to search this area properly," she said as she peered up six floors of elevator shaft.

They were, evidently, mall elevators.

"Can I have the card please?" Ayden said, peering around as his eyes adjusted to the gloom. "Maybe we can short circuit the process." He noticed that the agent waited for Henri to nod before handing him the card. Two of the elevator well walls were bare concrete but the other two were more compressed cellulose panels. He tried waving the card past a few likely looking panels with no result.

"We may need the scanner in here," he called out over the noise of the elevator nearest him, which was descending. One of the agents disappeared back down the corridor they'd entered by. Henri and the other agent spread out, examining every nook and cranny.

He walked to a ladder attached to the nearest wall and climbed it. From it, he stepped easily onto a small maintenance platform built into the roof of the now stationary elevator. Henri noticed, and walked toward the ladder.

"What's the idea, Ayden?"

He waved the card past some sort of maintenance panel, ignoring its buttons.

"I don't really know. I'm just going with the flow. How I'd organize things if I had set this all up. I think we probably need to go up."

As Ayden spoke, the elevator began ascending. Henri hadn't quite made it to the top of the ladder, the response team agent was just below her.

"Ma'am, that's against procedure and dangerous," the agent said.

"I know, dammit," she said under her breath, looking up at the departing elevator.

"Sorry," Ayden said looking down, smiling. "I'm just trying to move fast so he doesn't slip away again."

"So, why up? Come on, you have a specific idea, give!"

"Ventilation."

"Spill it, Ayden."

"In a box in a basement, he would quickly use up the oxygen," he called down. "Fill it up with carbon dioxide. Dangerous. So you either need a powered ventilation system, or you just put your box at the top of the building where there's endless natural ventilation."

Ayden wasn't sure she could still hear him but he heard her yell at the top of her lungs.

"Don't you damn well go anywhere, Ayden Walker! Dammit!"

. . .

He stepped off the elevator and headed down the lone corridor. Suddenly, his specs went opaque and Henri's avatar loomed large. It was entirely composed, of course, but he could hear the exertion in her voice. She was climbing an elevator well ladder the entire six floors.

"Ayden. I'm asking you. Please. Wait for us to get up to you. The DD'll string me up if anything happens to you."

"I won't run off on you, I promise."

Ayden quickly stripped off his gear and left it lying on the floor.

There was a door to the left which opened easily. Ayden stuck his head out. He was right at the end of another corridor but this one was wide and well lit. On the wall opposite was a sign pointing to public toilets. The mall again. He closed the door.

Looking back down the gloomy elevator well corridor, it obviously extended further than the public corridor beyond the door. Ayden examined the wall panels closely.

"Walker! Sir! Please wait! Please don't do anything!" Ayden heard the agent calling faintly. He was climbing the ladder too.

There. A panel which looked just slightly misplaced. The gaps were a little uneven.

"Mr. Walker! Sir?"

Ayden quickly swiped his card around the panel edges, listening to the clicks.

"The keys to the kingdom," he said quietly to himself.

The door sprang outward, catching Ayden by surprise. He staggered back, but nothing in the little room posed a threat. The panel had a door handle on the inside. Quickly he stepped inside, pulling it closed behind him.

. . .

Hanford was encased in a light sleeping bag. He lay on a thin mattress on top of two, large suitcases which were lying end to end on the floor. Transparent lines filled with fluid snaked from the suitcase under Hanford's head into the sleeping bag.

Intravenous drips, presumably. A small module was plugged into the suitcase which provided electronic readouts. Ayden crouched and glanced at them briefly. Heart rate, temperature, blood pressure, blood oxygen, and carbon dioxide partial pressures. The other suitcase was plain, revealing no signs of technology. Hanging on the wall above Hanford, incongruously normal, were hooks holding clothes bags. On the floor was a small bottle of water. The room was well ventilated: it was positively cold and draughty.

Ayden doubted there would be any booby traps as Hanford's body would be entirely vulnerable. Nevertheless, he didn't touch a thing, stood back again, near the door panel, trying to decide what to do next.

So it was a surprise when Hanford's eyes opened.

Ayden didn't speak. He simply watched as Hanford's eyes widened, taking in Ayden's presence.

Finally, Hanford was able to mumble. "Oh, Ayden, you fool. You clever, clever fool. You've ruined everything."

Hanford's hand was moving slowly within the sleeping bag. Shakily, it emerged. Ayden felt himself tense a little and then relax as Hanford wiped the grit from the corner of his eyes.

Then a finger slid toward a point behind his right ear. Ayden rushed forward, but he was too slow.

"Prometheus," Hanford said, barely audible, letting Ayden whip his hand away. His eyes never left Ayden's.

"You're alone?" Hanford said. He began to stretch, small, slow movements. It looked very uncomfortable. Even talking seemed an effort.

"Of course not. They're right behind me." On cue they heard Henri and the agent running along the corridor outside, calling for Ayden and shouting at each other and into their phones. He continued in hushed tones. "But they don't know where we are yet. Tell me."

"Why would I do that?"

"Because there's a chance I'd have agreed with you in the first place if you'd told me the whole story. I couldn't because I knew you were hiding something."

"Too risky."

"But now? The situation is very different now."

"What advantage?"

"You might convince me yet. At the least make me more sympathetic toward your decision to try to kill me."

"Sorry."

"Sorry you tried or sorry you failed?"

"Both."

"Tell me."

"Ahh." Hanford craned his neck. "You tell me. You know what needs to be done. You lack only the courage of your convictions to do it."

Ayden gave voice to the suspicion he had held for weeks. "Your virus causes a stronger immune system response in guys with higher baseline testosterone levels, strong enough that their own sperm are targeted."

Hanford said nothing.

"I'll confirm it anyway, in time," Ayden said. "You're targeting alpha males."

"That's imprecise terminology, but it'll do."

"You're taking the alphas... and would-be alphas, out of the next generation. How can you know the ultimate impact of such a radical change?"

"If you're facing a terrorist in control of a nuclear bomb and you get the chance to kill him, do you? Or do you worry about what the world might miss in the absence of his possible future children's contributions in fifty years time? It's the alphas who plunge their countries into war, who damage the environment irreversibly in order to increase their social status, who prey on their fellows for personal advantage. It's frustrated would-be alphas who brawl in bars, who rape, who kill. They place dominance above the lives of those around them, even above their own lives."

As he spoke, Hanford gestured toward the water bottle and slowly raised himself to a sitting position. He swung his legs down to the floor and leant back against the wall with his eyes closed. Ayden opened the bottle and handed it to him.

"They evolved for a reason," Ayden said.

"Of course. In some organisms, all males are effectively alpha." He took a swig from the water bottle. "Male bears and crocodiles lead isolated existences. The males fight on sight and the winners breed. The primary threat to a baby bear is its own father. The problem humanity faces is that modern technology makes our alphas too powerful, capable of too much destruction.

"In many organisms though, a second sexual strategy was chanced upon. Beta males, not driven to dominate, more cooperative, content to follow. In the immediate absence of the silverback, the females permit cooperative beta males reproductive access."

Hanford's ambition, his arrogance, was breathtaking.

"So you re-engineer humanity on the fly. The risks are… obscene."

"Are the risks likely to be worse than where we are currently headed? I think not. In any case, I had planned to be around for a long while to keep an eye on developments. Besides, it doesn't take all the alphas out of the gene pool. It just changes the balance. A balance that could be adjusted again, if necessary."

"You thought you could fine tune. Where did you expect leadership to come from?"

"Betas make great managers. They just need guidance to avoid becoming too bogged down in bureaucracy to achieve anything."

"With the ultimate leadership coming from you."

"And, in time, others like me, like us. Don't mistake me for a semi-aware homo sap alpha male, Ayden. I didn't do this for power or self-aggrandizement. I planned to influence from the background. I have no desire to make myself some sort of emperor."

"You planned eugenic apartheid."

"Apartheid was a form of oppression. I've been trying to stop oppression. Think of a small group of truly self-aware and rational leaders quietly guiding policy from behind the scenes, the great majority of people living their lives free of war and persecution and rape and murder."

"And free of self-determination."

"Self-determination is a red herring. Most people only really want a limited amount. What they want most is security, comfort, and social advancement."

"You keep saying 'they' and 'them.' You don't consider yourself part of humanity?"

"You and I are qualitatively different, Ayden. Technically, you and I are, of course, homo sapiens—we can still breed with them. But, perhaps it's time for a speciation event. Perhaps it's time for a new species to emerge, 'homo sapiens rationalis.' Rational man rather than merely intelligent man. Perhaps we represent the start of something fundamentally new. Envisage two sub-groups, call them subspecies if you like. Homo sapiens rationalis provides overall direction. Within that framework, homo sapiens beta has self-determination, limited only by the exhortation–do no harm."

"The description is utopian but that was ever the promise of the dictator."

"The promise of megalomaniac alpha males, yes, but you know I'm not that. Come on, Ayden. Imagine a generation not dominated by unthinking, self-absorbed aggression. Does that really sound like such a bad thing?"

"It sounds like fantasy. Contemplate my experience of you. Do you really think I could trust anything you say?"

"That's irrelevant now. Thanks to you, my capacity to influence events in the near future is likely very limited. Question now, is can you trust yourself to–"

Metal shrieked. The panel door exploded inwards.

Behind the combat-amour clad, crowbar wielding response team, Henri's face was grim.

Saturday, June 24th

"You're fretting unnecessarily," the deputy director said. "The situation's entirely contained. Hanford isn't going anywhere."

Ayden certainly had been concerned. For nearly two weeks he hadn't been able to get access to Henri or the deputy director. They had raced Hanford away from the mall in an ambulance and refused to tell Ayden what was going on. Henri dropped out of sight and all she would tell Ayden by encrypted message was that his entrance to the hide-away had triggered some sort of alarm, which had immediately administered a cocktail of wake-up drugs. The effects of Hanford's Prometheus command were still being determined.

Finally, the deputy director agreed to see Ayden.

"You did an extraordinary job, Walker. Without you, we wouldn't have Hanford. Thank you. I'm sorry we've had to keep you in the dark for a few days." He didn't sound sorry at all.

"May I ask where he is?" Ayden said.

"Somewhere familiar to you. He's in Colonel Soulandt's cave, but don't worry, we've upgraded security substantially since you were allowed to wander off."

Ayden resisted cautioning the deputy director against under-estimating Hanford. With limited egress to the surface, the cave could, no doubt, be made extremely secure. Clearly, however, it was more than a holding facility.

"Why there?"

"All samples and records of his vaccine were destroyed," the deputy director said.

"Prometheus?" Ayden said.

"Correct. It also launched a coordinated set of legal actions from firms around the globe. Suddenly we were surrounded by lawyers claiming they had proof that Hanford was alive and had been illegally detained, demanding access to their client. And the president had calls from a dozen congressmen."

"He's offered to re-create the vaccine for you."

"Yes. And to derive a cure for those already infected."

"And, no doubt, he's offered to re-create the longevity research he's been smothering, the records for which were also destroyed."

A brief, wintery smile crossed the deputy director's face. "That too," he said. "I would remind you, Walker, that if you'd waited for Soulandt and her team, they might have prevented the command. As it is, my hands are tied. As you've pointed out, we need that vaccine. And I can't just pretend we haven't found him. Equally though, he didn't much fancy half a century in jail, so we reached an accommodation."

"It's a dangerous compromise."

"It is. Welcome to my world, Walker."

"Which is why you agreed to see me."

The deputy director simply waited.

"You want me to help keep an eye on him, ensure he doesn't engineer some sort of program of his own from within the cave."

"I think we both want essentially the same thing, Walker."

"Actually, Deputy Director, what I want is to return to a normal life, free to explore the world through science and spend time with family and friends but, yes, I think I have a responsibility to keep an eye on him."

"Can't put the genie back in the bottle."

Ayden paused, considering. "But, that end wouldn't be best served by shutting me in there with him. I agree I should visit the cave to observe from time to time, but where I'll be most useful is out and about, looking for subtle indications of projects which continue despite his incarceration…or of other projects which may have been triggered by "Prometheus.""

The deputy director barked a laugh which held little humor. "So we have another 'Walker Special.' A solution which offers me the best likelihood of a positive outcome whilst, coincidentally, giving you free rein."

"Please give it serious thought. It's where I'll be most useful."

"Assume I do. Assume I and my people can't think of a better use for you. What then?"

"I take the role of roving investigator, keeping in touch with developments in the cave and visiting there regularly. I wouldn't require substantial resources and, of course, I'd report regularly."

"You most certainly would, young man. Hmmm, we'd need to provide you with an assistant, a facilitator to help you to access resources, as required."

"A minder. Of course. Let me guess. She'd be very intelligent and very attractive, and encouraged to get as close to me as possible."

The deputy director didn't respond but for the first time, his smile held true mirth.

"May I make a small suggestion?" Ayden said.

The deputy director inclined his head.

"I'm quite introspective and relatively objective in observing myself. I've also had quite a long time to get to know myself, so unless you think your psych agents are much smarter than me, please skip the games and just provide me with someone intelligent and intellectually open."

"I'll consider your proposal, Walker." The deputy director's face returned to its usual serious demeanor. "In the meantime, you are going to pay a visit to the cave. See if you can identify any security weaknesses, biological, electronic or otherwise. Colonel Soulandt has been supervising the security refit. She'll contact you. That's all for now."

• • •

After Ayden had left, the deputy director called his assistant into his office. "The team we've established to watch Walker…I'm assigning myself team leader for a while."

"But, sir…"

"Soulandt won't like it but she stays on as my second. Tell her to draw up revised responsibilities. Minimize my time but I make the decisions."

"Yes sir."

"Reschedule my diary accordingly. Daily reviews. Also, provide Walker with full access to me."

"Full, sir? How full?"

"Full. Per the President of the United States. Clear?"

Wilhelmina was dumbstruck. The deputy director gave her a moment to compose herself.

"Good Lord…what just happened in here?" she said, forgetting herself somewhat.

"Just what he wanted to happen, I suspect. I've got a couple of tigers by the tail here, Wil."

In nine years together, it was the first time the deputy director had called her by her first name.

"Sir, may I ask? Ayden Walker. Is he part of the solution or part of the problem?"

The deputy director sighed tiredly. "Both, I suspect."

Monday, June 26th

From the surface, the base was almost unrecognizable. In place of barely visible sensors on trees and in the ground, were three layers of fencing. The inner fence was a very fine mesh. It looked like it would keep small rodents out. Between the two outer fences, a pack of dogs patrolled, each animal wearing some sort of skull cap. Mounted on the outside of the outer fence at regular intervals were signs reading: "Department of the Environment. Closed ecosystem testing. No trespassing. Maximum penalty—$250,000 fine and seven years jail."

Inside the perimeter were several additional small, prefabricated buildings and a huge inflatable dome.

Henri met him inside the aircraft hanger. She looked tired.

"Jailer again, Henri?" Ayden said.

"Seems to be my lot in life, doesn't it," she said. The smile she gave him didn't reach her eyes. "Keeping watch over intractable geniuses. Actually, my brief is broader these days. Once this place is ship-shape I'll just drop in now and then to keep everyone on their toes."

"You've been busy."

"Yes. I'm largely done here for the moment but the DD tells me I'm keen for you to look over the place."

"I see. What've you done to the dogs?"

"Actually, less than you'd expect. The caps are really just camera and sensor mounts with little clickers built into them."

"Clickers?"

"Yes. Apparently, in the early days it was all much more invasive. Electrodes were surgically implanted into their brains. The idea was to control the dogs remotely. Basic stuff. Stop, go, left, right, look up, look down. So an operator could check things out using the camera. The control system worked but the surgical procedure was complex and the dogs were always fighting infection. Finally, someone realized that all they had to do was train the dogs to respond to some sort of remote communication, the way hunting dogs have done to whistles for millennia. Now, they are just fitted with a removable cap which produces a click only the wearer can hear."

"Keep it simple."

"Outer layers are more hi-tech though. A hundred surveillance micro-drones provide fine resolution coverage out to a mile or so. Continuous satellite surveillance for hundreds of miles."

Henri gave Ayden a thorough briefing of the additional precautions they'd taken. Really, there was nothing he could add.

"Of course, that just means he'll find more subtle means to achieve his goals. You don't rotate staff here frequently do you?"

"Can't. That'd widen the knowledge circle too much. Besides, this is still a research facility. We'd lose too much expertise."

"So one approach will be to try to influence the staff, particularly the other scientists. He'll be patient and subtle."

Another thought occurred to Ayden. "Make sure you thoroughly incinerate all lab waste."

"We do anyway. You think he might try to get something out that way, then convince a scientist to activate it on the outside?"

"It's possible. Or he could cook up a virus, and infect everyone here with it. Your people then become the delivery agents to the world outside."

"Okay, so…we do a full-blood DNA genome sequence for all staff on entry and exit. Parasites, pathogens, nuclear DNA, the lot. If someone gets a virus in here, we'll know about it."

"Is that workable?"

Henri grimaced. "Probably mean a twenty-four hour quarantine. Won't be popular but, other than me, no one gets in or out of here without advance notice anyway."

"Are you talking about the entire genome or just samples?"

"All three billion human base pairs plus about another billion belonging to the microscopic ecosystem we all carry around. Sequenced and analyzed."

"That's fast. Hmm…equally important I think, will be making sure you really understand everything he works on."

"We may need your help on that."

"No problem. I imagine you have a full electronic blackout on him?"

"Essentially. You know the drill, re-keyed text, some re-authoring to make hidden codes more difficult."

Their review of the security protocols finally wore down.

"Can I see him?" Ayden said.

Henri hesitated. "The deputy director has given you clearance." She lapsed into silence.

Ayden waited.

"I argued against giving you access. Mainly because I never know what you're damn well going to do next and partly because, well…it won't serve any real purpose. Look, he tried to kill you. He nearly succeeded. What'll it achieve to talk to him?"

"This may sound strange to you, Henri, but in a way, I'm okay with his attempt to kill me. I believe he really didn't want to do it. That's why he took the risk of leaving with the job incomplete—he didn't want to see it. But, I put him in the position where he had to make a difficult decision to sacrifice me for what he sees as being the greater good."

She stared at him as though she was suddenly having difficulty recognizing him.

"*May* sound strange? Don't you make excuses for that sonofabitch, Ayden. He tried to kill you, and he tried to enslave the human race. He's a goddamn psychopath and if there was ever an argument for capital punishment—or state-sponsored assassination if it comes to that—then as far as I'm concerned, he's it. If they need someone to pull the trigger, I'll put my hand up."

The silence that followed was uneasy.

Finally, Ayden spoke. "We really are very close to sending our planet into a tail-spin. Whether you accept or reject his approach, Henri, he was trying to save the world."

"Do you realize how much you isolate yourself when you say stuff like that?"

Suddenly, she was standing. Something in her face seemed to close.

It was time for Ayden to talk to Hanford.

• • •

Inside the base, the changes were less apparent, though Ayden noticed that Mother had gotten her fake grass. They sat in the main cave space, on a bench beside one of the large planter boxes. They could have booked one of the meeting rooms, but they were both acutely aware that any semblance of privacy would be purely illusory.

They shared the bench, each sitting comfortably still, looking ahead rather than at the other. It would have been easy for a passer-by to assume they were ignoring each other.

"I have to admit to admiration, Ayden. You played me. That hasn't happened since I was a child. You've also done something terrible for which you will now pay the price."

Ayden didn't speak. He simply waited for Hanford to continue.

"You thought I was the problem. As such, catching and imprisoning me was the solution. That was childish. Now, you'll be forced to grow up. In keeping with the demonic role you've assigned me, I've cursed you."

Hanford paused again. This time the pause stretched. Ayden's peripheral vision made him turn his head. Hanford was now looking directly at Ayden. Ayden examined his smile. It wasn't triumphant or predatory. Was it ironic? Perhaps even sympathetic?

"I've cursed you with knowledge. Knowledge you won't be able to turn your back on. These people," Hanford gestured broadly. "Are delivering mortal blows to the biosphere. It will ultimately survive, and so will humanity, but only after suffering and death on an unprecedented scale. I had assumed the burden of averting that disaster, but you took it upon yourself to stop me. Now, it's yours. Stopping me was the easy bit. Now, you have the responsibility to find a solution which fits within your sensibilities."

Hanford stood, looking straight ahead.

"Come back and talk to me again when you've grown up," he said walking away without a backwards glance.

• • •

The government jet was luxurious, dominated by plush carpet and deep leather armchairs. He was offered tantalizing morsels of food by a beautiful hostess. Ayden was on his way to Washington, to a life of influence and comfort, if not privacy. He had negotiated a reasonable degree of freedom and access to substantial resources. He would even be free to spend time with his parents and friends.

But his every interaction with them would be observed and analyzed. He would not inflict that on his mother for more than brief visits.

He'd alienated Tess.

He'd even managed to distance Henri. Their parting had been perfunctory. She hadn't really met his eyes since he'd admitted sympathy for Hanford's approach.

And Hanford was right, at least in part. The vaccinations were proceeding. Hundreds of millions of men had been infected, but a cure was being developed and the rest would reproduce normally and their children would all demand the highest possible standard of living.

Which meant something had to be done. Urgently.

Ayden *did* feel responsible to come up with a solution.

But he didn't have a solution. He couldn't think of anything even remotely morally acceptable which had a chance of doing enough, quickly enough.

The responsibility was crushing. His life had been a specious fantasy, cocooned by an ignorance he was now denied. He mourned its passing. Again.

The small aircraft jostled its way through clouds as it began its descent. Rain lashed at the window. Ayden peered out at the endless shades of grey.

● ● ●

They had landed. The jet's crew was waiting for him. He physically shook himself. He'd been feeling sorry for himself. That was unacceptable. He'd led a fortunate, even privileged, life.

Ayden descended the stairs and climbed into the limousine waiting to take him to the deputy director.

15:
JULY 2034

He couldn't entirely hide his age. Make-up almost hid the liver spots on his face but it couldn't hide the wrinkles or the sagging skin. His hair was a perfectly uniform rich brown but exceedingly thin. He didn't try to hide his fury.

"In an outrageous move today, Congress passed its indefensible Old Age amendment to the Social Security Act so even if you've been contributing for fifty years through your payroll taxes you'll no longer be able to claim any benefits 'til you reach sixty-eight."

A tiny dribble of spittle escaped from the corner of his mouth.

"Republican Senator Thomas Lim tried to defend the legislation today saying it's a vital part of the austerity package forced on his government, in part, because our generation waited so long to act on climate change! But how were we to know it'd be this bad or cost this much? The science was...there were uncertainties! It was very complicated!"

Gresham Cain Today, "Senior Citizen Spokesman Calls For Senior Voters to Write to Their Local Congressman in Protest, as the Attainment Age for Retirement Insurance Benefits Is Raised from Sixty-Two to Sixty-Eight," today.seniors.us

Sunday, July 16th

Tess had sent Ayden a brief text message. As instructed, he rented an all-terrain vehicle and picked her up from her hotel wearing hiking gear.

Soon they were walking up the south spur of Mount Tamalpais. They talked about the weather, about family, about the progress of the worldwide vaccination program. It felt great to be talking to her again. Shortly after noon, they found a comfortable spot to rest and eat their trail food.

"Ayden…I was wrong," Tess said, then paused as Ayden held up a hand.

"Before you continue, Tess, I have to warn you. Even up here, we may not have true privacy. There could well be binoculars and a sniper mic trained on us right now." He looked around the hillside, not really trying to detect surveillance attempts. "Someone will be listening at least. I'm sorry."

"Well, I'm just going to have to get in touch with my inner exhibitionist." Tess stood with a fierce grin. She faced generally down the valley, then hauled up her sweater together with the t-shirt and bra underneath, flashing her breasts at the countryside.

"What—"

"The hell with them," she said, restoring her clothing and sitting down beside him. She smiled demurely. "Close your mouth, Ayden, you'll catch flies."

Their grins broke into giggles. Staring at each other, it grew. Rolling on the ground holding his belly with laughter, Ayden felt a landslide of release within himself. They lay for a while, quietly gazing up at the sky.

Finally, Ayden broke the silence. He turned his head to look at her. "You know the main difference between William and I, Tess?"

She propped herself up on an elbow so she could look him in the eyes. "What?" she said.

"You. You and my parents. From the perspective of a purely rational outsider, a 'disinterested party' if you like, it's hard to argue with William's approach to managing homo sapiens."

"So why'd you stop him then?"

"Well, I guess I've always been a bit of a loner. I was never really one to 'run with the pack,' but because of my relationship with Mom and Dad, and because of you, I don't consider myself an outsider. Unlike William, I don't see humanity as just another species to manage."

She studied him.

"You know, I think I finally understand my father, Ayden," she said. "I've always found his obsession with work maddening but he always says to me that I need to decide, not what to *do*, but what to *be*, to keep looking till I find it, then to be it with everything I have. He says he pities those who thrash about, endlessly seeking distraction from their own lack of purpose."

She sat up, cross-legged in the thin grass, her eyes still locked on his.

"Well, I've done a lot of thinking over the last few weeks and I've decided. I can't ignore the mess humanity is making of the planet. I want to help you fix it…and I'll put up with inertia and ignorance and invasion of my privacy. We'll find solutions, and they won't be perfect but I won't beat myself up about that and I won't let you beat yourself up about it either."

"Wow," Ayden said. "I don't deserve you, Tess."

"Oh, yes you do. But."

"Yes?"

She said it with a smile. "No more of this keeping me in the dark crap. Okay?"

"Ouch. Yes, okay. But, sometimes it takes courage to *hear* the truth, doesn't it?"

Tess winced. "Touché."

There was silence for a while.

"Let's walk," Ayden said.

He filled his lungs and looked around. The sky was a deep, vivid blue, the hillside a cacophony of greens.

They walked along the ridge side-by-side, in the crisp clarity of the mountain air.

THE END

. . .

If you liked INTERVENTION, please consider posting a brief review. Reviews are lifeblood for independent authors.

COMING SOON!

If you would like to know what Ayden and Tess face next, keep a lookout for the sequel,

INTEGRATION

If you'd like to know when INTEGRATION will be available, visit
integrationbymunro.com

If you'd like to know about the Author, visit
wrrmunro.com

W.R.R. MUNRO

ACKNOWLEDGEMENTS

If I'd known what I was getting myself into, I'm really not sure I would have started. What began as an idle hobby became an obsessive joy. Until I finished the first draft. Then it became hard work, learning the basics of the craft of writing. Rewriting, rewriting, rewriting.

Among my many foolish misconceptions was the idea that I could just sit down and produce something readable by myself. Not so. Oh, so much not so.

Without the patient, persistent advice and encouragement of my sister, Emma Munro, there's no way this book would have gotten past its clumsy, inept first draft. (Thanks Em, you're up!)

Many thanks also to my 'beta' readers; Yael Grinham, Jamie Everingham, Mark Woodard, Belinda Stafford and Jackie Harris.

Thanks also to Sean Doyle of Lynk Manuscripts (lynk.com. au), and to Erica Orloff of Editing for Authors (editingforauthors. com). And thanks to Kris Kendall of Final Edits (final-edits.com), and to Amy and Rob at 52 Novels (52novels.com) and to Michael Campbell of MC Writing (mcwriting.com). And finally, thanks to Jeroen Ten Berge (jeroentenberge.com).

If you liked the book, it's in no small part due to the help of the above. If you didn't, blame me solely.